Information Systems Engineering and Management

Volume 25

The book series "Information Systems Engineering and Management" (ISEM) publishes innovative and original works in the various areas of planning, development, implementation, and management of information systems and technologies by enterprises, citizens, and society for the improvement of the socio-economic environment.

The series is multidisciplinary, focusing on technological, organizational, and social domains of information systems engineering and management. Manuscripts published in this book series focus on relevant problems and research in the planning, analysis, design, implementation, exploration, and management of all types of information systems and technologies. The series contains monographs, lecture notes, edited volumes, pedagogical and technical books as well as proceedings volumes.

Some topics/keywords to be considered in the ISEM book series are, but not limited to: Information Systems Planning; Information Systems Development; Exploration of Information Systems; Management of Information Systems; Blockchain Technology; Cloud Computing; Artificial Intelligence (AI) and Machine Learning; Big Data Analytics; Multimedia Systems; Computer Networks, Mobility and Pervasive Systems; IT Security, Ethics and Privacy; Cybersecurity; Digital Platforms and Services; Requirements Engineering; Software Engineering; Process and Knowledge Engineering; Security and Privacy Engineering, Autonomous Robotics; Human-Computer Interaction; Marketing and Information; Tourism and Information; Finance and Value; Decisions and Risk; Innovation and Projects; Strategy and People.

Indexed by Google Scholar. All books published in the series are submitted for consideration in the Web of Science.

For book or proceedings proposals please contact Alvaro Rocha (amrrocha@gmail.com).

Anastasiia Mazurchenko

The Impact of the Digitalization on the Human Resource Management

Springer

Anastasiia Mazurchenko
Department of Business Administration
and Management, Faculty of Economics
Technical University of Liberec
Liberec, Czech Republic

ISSN 3004-958X ISSN 3004-9598 (electronic)
Information Systems Engineering and Management
ISBN 978-3-031-76789-0 ISBN 978-3-031-76790-6 (eBook)
https://doi.org/10.1007/978-3-031-76790-6

This Springer imprint is published by the registered company Springer Nature Switzerland AG
The registered company address is: Gewerbestrasse 11, 6330 Cham, Switzerland

The book is dedicated to the Author's family, who is going through a war in Ukraine.

Acknowledgments

The Author wants to thank her dissertation supervisor, Assoc. Prof. Dr. Kateřina Maršíková, for her professional guidance, valuable advice, motivation, time, and patience since the dissertation is the basis for this book. Special thanks also go to all the members of the Department of Business Administration (DBA) and the Head of the Department of Economic Statistics of the Faculty of Economics, Technical University of Liberec (FE TUL), Dr. Vladimíra Hovorková Valentová, with their professional advice and support, who were helpful in the preparation of the empirical part of the surveys. The Author also wishes to thank all respondents for their willingness and cooperation in participating in the surveys, the results of which are used in the book. Last but not least, the Author thanks her parents and husband for their help, patience, and support during the preparation of this book.

About This Book

The book is divided into seven main chapters. In Chap. 1, attention is paid to determining the main objective and sub-objectives (SO) of the book, formulating research questions (RQs) and hypotheses, and evaluating them, which is the main output of the book. Chapter 1 also summarizes information on the methodology of the empirical part of the book, the sources of primary data, and the research methods used in their analysis and evaluation.

Chapter 2 presents the results of a systematic review of literature sources dealing with the historical development and definition of digitalization under the book's defined theoretical framework. Chapter 2 also describes the impact of the COVID-19 pandemic on digitalization in corporate practice.

Chapter 3 presents the historical development and emergence of digital human resource management (HRM) concerning the industrial revolutions that have occurred before and the Fourth and Fifth Industrial Revolutions currently underway. Chapter 3 also considers the current trends in digital HRM and their main advantages and disadvantages.

Chapter 4 defines the role of digital competencies and their use in competency models in the Fourth Industrial Revolution for HRM purposes.

Chapter 5 presents selected secondary data from global studies on the impact of digitalization on the labor market, HRM, and the competency requirements of employees and HRM practitioners. At the same time, Chap. 5 justifies the selection of particular sectors of the economy for further investigation of the impact of digitalization on HRM in the empirical part of the present book based on the secondary data search.

Chapter 6 is devoted to evaluating the attitudes of employers and employees operating in selected sectors in the Czech Republic (CR) and abroad towards digitalization and digital competencies. Chapter 6 also evaluates the results of empirical research carried out by the book's Author about the ROs and hypotheses.

Chapter 7 summarizes the book's primary results, limitations, and contributions in theoretical and practical terms and defines the possible direction of future research. It also proposes recommendations for employers based on the results of the empirical research conducted.

Contents

Abbreviations

ACEA	European Automobile Manufacturers Association
AI	Artificial Intelligence
ANOVA	Simple Analysis of Variance
ASME CR	Association of Small and Medium-Sized Enterprises and Crafts of the Czech Republic
CAS	Czech Agency for Standardization
CAWI	Computer Assisted Web Interviewing
CEOs	Chief Executive Officers
CITB	Construction Industry Training Board
CLC	Construction Leadership Council
CMCTU	Czech-Moravian Confederation of Trade Unions
CR	Czech Republic
CS	Combined Studies
DBA	Department of Business Administration
DCC	District Chamber of Commerce
DigComp	Digital Competence Framework for Citizens
DT	Digital Technology
DTs	Digital Technologies
EC	European Commission
e-HRM	Electronic Human Resource Management
ENIAC	Electronic Numerical Integrator and Computer
EU	European Union
FE TUL	Faculty of Economics, Technical University of Liberec
FGDs	Four Focus Groups
HEI	Higher Education Institution
HEIs	Higher Education Institutions
HR	Human Resource
HRIS	Human Resource Information System
HRM	Human Resource Management
HRs	Human Resources
IAEA	International Atomic Energy Agency

ICDL	International Computer Driving Licence
ICT	Information and Communication Technology
ILO	International Labor Organization
IoT	Internet of Things
IT	Information Technology
LLC	Limited Liability Company
MS	Microsoft
NPI CR	National Pedagogical Institute of the Czech Republic
NSO	National System of Occupations
OECD	Organization for Economic Co-operation and Development
PIAAC	Programme for the International Assessment of Adult Competencies
RQ	Research Question
RQs	Research Questions
SMACIT	Social, Mobile, Analytics, Cloud, Internet of Things
SMEs	Small and Medium-Sized Enterprises
SO	Sub-Objective
SOs	Sub-Objectives
T&D	Training and Development
UK	United Kingdom
UNIVAC	Universal Automatic Computer
US	United States
UTAUT	Unified Theory of Acceptance and Use of Technology
VR	Virtual Reality
WEF	World Economic Forum
WoS	Web of Science

Chapter 1
Fundamentals of Researches, Methodology, and Solution Procedure

Abstract Chapter 1 presents the book's primary objective and sub-objectives, contributing significantly to its fulfillment. It also poses the research questions, which are linked to the hypotheses. Furthermore, the conceptual and theoretical framework of the book is presented. Chapter 1 details the research procedure and methodology to fulfill the book's objectives. The basis for analyzing primary and secondary data in the book is a systematic literature review of Czech and foreign sources related to digitalization and its impact on human resource management. For the book, general theoretical research methods were used, including analysis, synthesis, deduction, comparison, and specific methods, which include quantitative methods associated with the hypothetico-deductive approach (i.e., statistical tests) and qualitative methods (i.e., individual in-depth interviews and case studies).

1.1 Research Objectives

The main objective of the book is to determine the intensity of the impact of digitalization on selected areas of HRM, with an emphasis on the use of digital competencies in the context of the COVID-19 pandemic from the perspective of employers and employees, based on an assessment of the current state of digitalization in the context of human resources (HRs) in selected Czech and foreign organizations and to map these competencies concerning competency models in organizations.

SO_1 identifies critical areas for organizational learning and development in terms of digital competencies according to the views of employers and employees. SO_2 identifies which competencies are necessary for using digital technologies (DTs) at work and how requirements differ across sectors. SO_3 also checks whether the competence profiles of existing employees are in line with employers' requirements regarding what digital tools are currently used in organizations and how employers support their employees in learning and developing digital competencies.

The assessment of the current state of digitalization in HRM was based on:

- Systematic literature review, the results of which are presented in Chaps. 2, 3, and 4;

A. Mazurchenko, *The Impact of the Digitalization on the Human Resource Management*, Information Systems Engineering and Management 25,
https://doi.org/10.1007/978-3-031-76790-6_1

- Secondary data analysis presented in Chap. 5;
- Primary data analysis, the main outputs of which are presented in Chap. 6.

The impact of digitalization on HRs in organizations and the level of the digital readiness of current employers and employees were examined from two perspectives in the primary data collection and analysis. On the one hand, employers' attitudes from Czech organizations were presented in selected sectors towards digitalization and its impact on business practice. It included mainly companies operating in the following sectors: banking and insurance, construction, and manufacturing, emphasizing the automotive industry. The rationale for the selection of the sectors is presented in Sect. 6.2. Conversely, the employees' opinions on developing their digital competencies at work were obtained from a sample of respondents operating in different sectors. These respondents were employed students of the combined form of study (CS) at FE TUL. Here, the attitude of employees of two selected higher education institutions (HEIs) in the CR and Germany was also considered.

The research was carried out in several phases:

- The first phase involves a pilot survey and then a comprehensive questionnaire survey of Czech businesses operating in the selected sectors.
- The second phase is based on conducting in-depth interviews with selected employees responsible for HRM in cooperating Czech enterprises.
- The third phase focuses on collecting primary data from employees working mainly in the Czech and foreign academic environments and other sectors of the Czech economy.
- In the final phase, the results are analyzed. The results obtained at a HEI in Germany were presented and submitted to its management as a final report.

In order to achieve the main objective and sub-objectives of the book, it is necessary to define and implement the following steps, which are in line with the focus and structure of this book:

Step 1: Defining the theoretical framework of digitalization and its impact on HRM based on a systematic literature review of professional Czech and foreign publications.

Step 2: Preparing the database of Czech companies operating in the CR based on data from the Bisnode MagnusWeb database according to selected criteria. The criteria for preparing a functional data set for research activities include an active legal entity, active company, known number of employees, exclusion of cases of being in bankruptcy or liquidation, absence of significant adverse events in the company's history, the existence of a functional company website, and availability of an active e-mail address in the contact details section.

Step 3: Selection of a research sample from the database of Czech companies based on a search of current knowledge on the impact of digitalization on human resource (HR) development in selected industries.

Step 4: Primary data collection using quantitative and qualitative methods supported by collaborating businesses.

Step 5: Applying the chosen methodological approach in practice, testing hypotheses and ROs, and evaluating the impact of digitalization on business practice and employee competencies from the employer's perspective.
Step 6: The research sample will be selected from employees working in Czech and foreign educational institutions and other sectors of the Czech economy. Primary data will be collected using qualitative methods. The employees will be assessed on the current status and possibilities for further development of digital competencies at work.
Step 7: Based on the primary data collected and secondary data, identifying critical areas for organizational learning and competence development from the perspectives of employers and employees in light of the ongoing digitalization accelerated by the COVID-19 pandemic.
Step 8: Synthesis and generalization of the findings and formulation of recommendations for employers.

1.2 Research Questions and Hypotheses

In line with the SOs defined in Sect. 1.1, based on the literature review presented in Chaps. 2, 3, and 4, RQ are formulated to provide relevant information for subsequent hypothesis testing:

RQ_1: What competencies do employers and employees think are necessary to effectively use digital technologies at work ?

RO_1 is aimed at comparing the views of employers and employees on the desired competency profiles of employees, given the increasing speed at which new DTs are being developed and introduced into organizations. A part of the professional community can find that the lack of digital competencies among employees is one of the reasons why most companies are forced to invest much more in the training and development (T&D) of their HRs. On the other hand, the dynamic development of digital technology (DT) makes systematic mapping of digital employees' competencies and their development to a large extent difficult, as the relevance of individual activities and associated competencies is constantly changing. Very few research studies address this issue and take into account a comparison of views between employees and employers. The answer to the above research question (RO) will be obtained through empirical research on the impact of digitalization on the competencies of their employees at selected Czech and German educational institutions.

RQ_2: How do priority digital competencies differ across sectors?

During the COVID-19 pandemic, more systematic surveys need to map the digital competence requirements of employees working in different sectors (further verified and discussed based on a systematic literature review). There is a debate among the professional community about whether digitalization has a different degree of impact on the competence requirements of employees about their age, the length of their employment, and the employer's sector of operation. Several researchers believe that all employers require specific digital competencies in the digital economy. RQ_2 aims to find out whether there are differences in which employers prefer digital competencies for the specificities of different economic sectors. The aim is also to check whether the age of employees or the length of time they have worked in a particular economic sector impacts their level of digital competencies. In order to answer the above RQ_2, the testing of statistical hypotheses H_1, H_4, and H_6 will be used.

RQ_3: How often and how do employees acquire the digital competencies they need for their jobs?

The answer to RQ_3 will be derived from empirical research on the digital readiness of employees in the banking and insurance, construction, and manufacturing sectors, emphasizing the automotive industry and the higher education sector in the CR and abroad. Statistical hypothesis testing H_2 and H_5 will also be used to answer RQ_3.

RQ_4: What are the main barriers to developing digital competencies at work from the perspective of employees and employers?

In determining the RQ_4, the Author of this book based on the assumption that the definition of barriers to developing digital competencies at work may differ between employers and employees. This topic has not received sufficient attention in the academic literature to date. RQ_4 examines whether employers and employees perceive various barriers to developing digital skills as significant and, if so, to develop recommendations for employers that will contribute to their removal. An empirical research investigation of digital competencies in corporate practice from the perspective of employees and digital readiness from the perspective of employees in the CR is being conducted to answer the above RQ_4. At the same time, the perception of barriers caused by digitalization in the work of university academics at selected Czech and German universities is compared.

RQ_5: How does digitalization affect organizational learning, development, and competency models applied in organizations?

Four in-depth interviews with HR specialists working in three selected Czech manufacturing companies and one bank will provide the answer to RQ_5. The results of the in-depth interviews will also be compiled into case studies to answer RQ_3.

The RQs will be answered using quantitative questionnaire surveys and semi-structured in-depth interviews with representatives of selected Czech business entities responsible for HRM. Based on the formulated objectives of the book, hypotheses were established, the confirmation or rejection of which will be presented in this book in Sects. 6.1.5 and 6.2.3.

> H_1: Employee competence level in using DTs is independent of the sector in which the employer operates according to the CZ-NACE classification of economic activities.

H_2: The frequency of improving employee competencies necessary for practical work in the digital environment is independent of the sector in which the employer operates, according to the CZ-NACE classification of economic activities.
H_3: According to the CZ-NACE classification of economic activities, enterprises' readiness to use DTs in the COVID-19 pandemic is independent of the sector.
H_4: Employees' level of competence in using DTs at work is independent of their age.
H_5: The frequency of improving employees' competencies in using DTs at work is independent of their age.
H_6: Employee competence level in using DTs is independent of their length of employment in the company.
Research hypotheses H_1, H_2, H_3, H_4, H_5, and H_6 are tested using selected statistical methods, which are presented in more detail in Sect. 1.7.

1.3 Research Concept and Methodology

The book consists of seven parts. The general procedure for preparing the theoretical and empirical parts of the book is presented below (see Figs. 1.1 and 1.2).

In the book's introductory part, the SOs of the book were first specified, the RQs were set, and the hypotheses were formulated. The basis for the elaboration of the book is the development of a systematic literature review of Czech and foreign scientific sources and the elaboration of theoretical knowledge in the field of digitalization in enterprises, its influence on the development of competencies of current employees, and the presentation of a competency model, which serves to fulfill Step 1. The selection of methods for implementing empirical research and collecting primary data also contribute to this.

The empirical part includes several sub-phases (see Fig. 1.2). Step 1, a pilot survey, was conducted on employers' attitudes toward digitalization in Czech enterprises. After testing and modifying the questionnaire, the MagnusWeb database of Czech enterprises was prepared, and respondents were selected to participate in further research, thus implementing Step 2. Next, a quantitative research survey was carried out using the modified questionnaire by electronic questioning method in the sectors identified from the previous secondary data analysis. In order to support the results obtained and reflect different ways of understanding the research problem, qualitative research was also carried out in the form of several in-depth interviews with HR professionals in the selected companies. Through these in-depth interviews, Step 3 will be fulfilled.

In the empirical part, an evaluation of employees' attitudes towards the topic under study is carried out, which includes the preparation, implementation, and analysis of the results obtained by the survey method on an online form in three focus groups. The research sample includes employees (data obtained on a group of employed students of CS at FE TUL), employees of a selected Czech university, and a German university. Step 4 compares the findings obtained from all three focus groups,

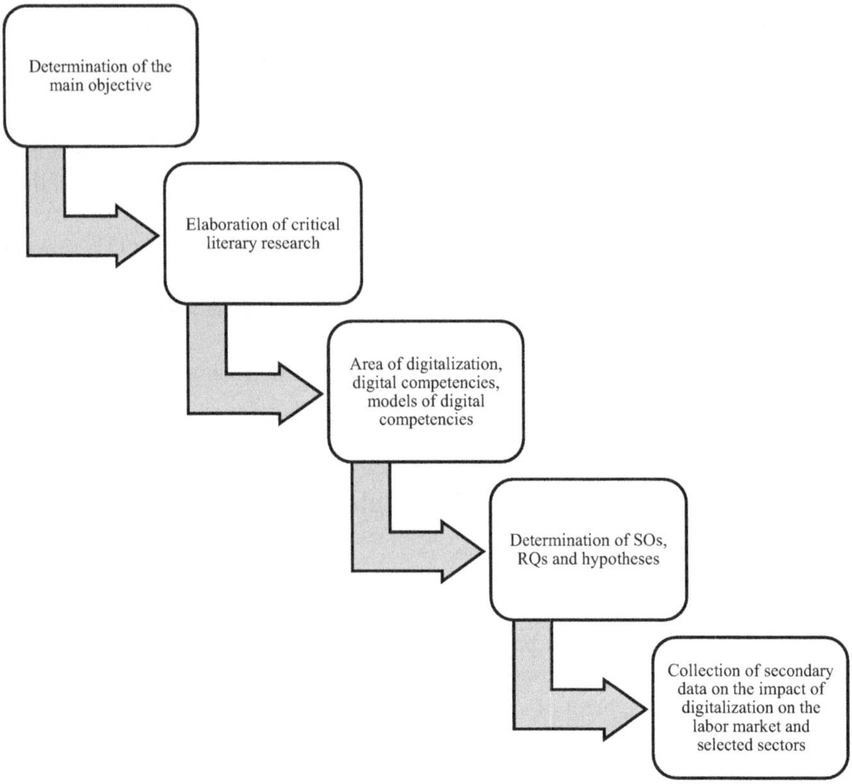

Fig. 1.1 Concept of the book development: Theoretical part

including interpretation of results, formulation of conclusions, and recommendations for employers.

The next part of the book is devoted to the main objective, i.e., research on the impact of digitalization on HRM in companies with an emphasis on the current period and the impact of the COVID-19 pandemic and the state of use of the model of application of employees' digital competencies in Czech companies. Thus, Step 5 is fulfilled. In the book's final part, the findings regarding the acceptance or rejection of the set hypotheses are presented. Based on the summary of the primary and secondary research results, conclusions are formulated, and Steps 6, Step 7, and Step 8 are fulfilled.

The following methods were mainly used for the elaboration of the book:

- For the preparation of the literature review:

  Monitoring and analysis of scientific literature from the Web of Science (WoS) and Scopus databases (see Sect. 1.4);
  Study and analysis of secondary data;
  Theoretical methods of synthesis and deduction;

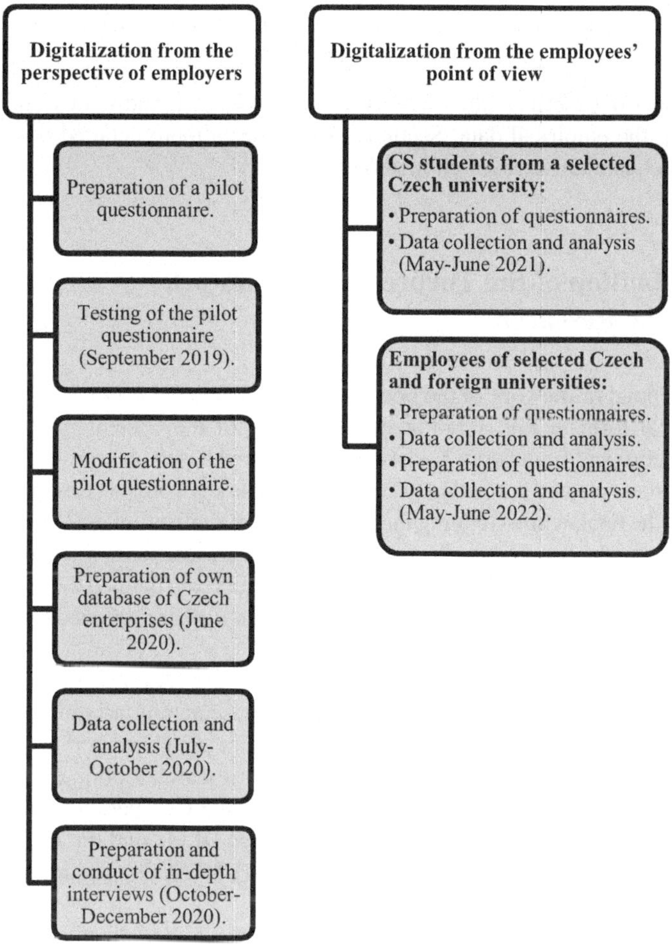

Fig. 1.2 Concept of the book development: Empirical part

Comparison of the findings.

- For the empirical part:

  Pilot survey (i.e., conducting a pilot survey on a randomly selected sample of respondents);
  Electronic questionnaire survey using the Computer Assisted Web Interviewing (CAWI) method using Survio survey software;
  Statistical tests for hypothesis testing, namely Pearson's chi-square test of independence, Fisher's exact test, Mann–Whitney U test, test of independence of ordinal correlation, simple analysis of variance (ANOVA), Kruskal–Wallis H test and Welch's test;
  Semi-structured in-depth interview (i.e., interview);

A case study of three Czech automotive companies and a large Czech bank on the impact of digitalization during the COVID-19 pandemic.

Section 1.6 describes the reasons for choosing the individual methods for processing the empirical data. Section 1.7 provides a more detailed description of the chosen quantitative research methods.

1.4 Definition of the Theoretical Framework of the Research

The conceptual framework of the book is presented in detail in Sect. 1.3. It is shown in graphical form in Fig. 1.1. For greater clarity of the content presented in the book, a theoretical framework has also been prepared based on segmenting the used expert sources selected according to keywords. These keywords were identified by the Author of the book while studying the literature on digitalization in HRM due to their importance for the comprehensive presentation of the researched topic. Subsequently, an overview of the publications of the principal authors who defined these key terms is also provided, including their publication dates with further reference to the sections in which they are analyzed in more detail.

The selection of suitable literature sources was primarily done in the electronically renowned scientific databases WoS and Scopus by searching according to selected keywords and their combinations using the operators "and" and "or". Other sources of information include academic monographs and research reports of international organizations such as the Organization for Economic Co-operation and Development (OECD), the International Labor Organisation (ILO), and the United Nations Educational, Scientific and Cultural Organisation (UNESCO). The rationale for the choice of keywords is given below.

The first keyword is digitalization. Definitions of digitalization are given in Sect. 2.2. It is a long-term phenomenon affecting more and more sectors and penetrating various areas of business and management. Digitalization refers to a set of concepts and definitions that characterize the impact of DT on the quality and productivity of production processes, the technical aspects of business development, the changing business model, organizational structures, and HRM in different sectors. The success of digitalization is closely linked to the positive net benefits for companies in the long term, which forces companies to face new opportunities and challenges.

To better illustrate the methodology of searching scientific publications for the theoretical part of the book, the procedure is presented below with the example of using the keyword "digitalization" when working with the WoS database in Table 1.1.

In the literature, it is also possible to come across the following key terms, sometimes used as a synonym for digitalization or as its next development phase. It is digital transformation. It is a complex interdisciplinary area closely linked to the changing expectations of employees, customers, and business partners and the conditions for

Table 1.1 Literature review process as of 31 August 2022

Search phase	Number of records in WoS	Categories of records in search results
1	17,548	Records were identified by searching with the keywords "digitalisation" or "digitalization"
2	15,735	Records after examining titles, abstracts, and keywords for "digitalisation" or "digitalization"
3	14,195	Records published in English, Czech, Polish, Ukrainian, and Russian (i.e., in languages that the book's Author knows)
4	13,045	Records are published in articles, conference papers, and chapters in books and scientific monographs
5	5,431	Records that are in open access
6	1,180	Records that fall under the research area "business economics"
7	486	Records that fall under the category of "management" in WoS
8	20	Records after reading the title, abstract, and methodology
9	7	Records closing the definition of digitalization

doing business under the influence of DT. From this perspective, digital transformation requires demanding preparation and readiness for risky decision-making, especially from top management. The difference between digitalization and digital transformation is explained in Sect. 2.2.

Another critical concept is DT, through which digitalization simplifies business processes, increases labor productivity, reduces operating costs, and adapts to market changes. In addition, DTs are also instrumental tools for attracting and retaining qualified employees, which is one of the essential objectives within the HRM strategy. DT is discussed in more detail in Sect. 2.3. It is important to emphasize that DT requires not only an understanding of their work but also a change in the mindset of employees to recognize and maximize the capabilities that these technologies require in terms of knowledge sharing, ways of communicating, managing business relationships, and improving business competitiveness. Last but not least, DT is also transforming HRM processes, the structure and functions of the HR department, and the activities of the respective employees, ultimately resulting in the emergence of digital HRM. This trend has revolutionized the nature of work and created a demand for a new set of competencies both on the part of HR managers and ordinary employees. The definition of the concept of digital HRM is discussed in Sect. 3.2.

Another essential concept chosen for the research part of the book is the competency model. It is because companies' efforts to adapt to digitalization lead to redefining the current competencies of employees and the competencies required of them in the future, affecting the objectives, content, and focus of training. The definitions of the competency model are given in Sect. 4.2. The competency model refers to a set of specific competencies, knowledge, skills, and behavioral requirements of employees that enable effective job performance for each job within the organization. The competency model can vary substantially from sector to sector and is usually driven by the needs and objectives of a particular company or organization.

Digitalization strengthens the role of selected competencies that are very important for practical work with DT and tools in the workplace. This justifies the need to ensure the continuous development of these competencies and their integration into existing competency models in companies. These digital competencies are simultaneously required to a greater or lesser extent in almost all professions and sectors. Digital competencies are presented in more detail in Sect. 4.2.2. To thrive in an increasingly digitalized work environment is a challenge that HR managers have had to deal with for a long time.

Table 1.2 presents a theoretical framework summarizing an overview of the prominent professional publications for this book's literature section. Based on a systematic literature review, the framework was developed gradually over six months.

Based on the SOs and the RQs presented above, the following chapters of this book will first conduct a systematic literature review on digitalization and its impact on HRM in organizations. Subsequently, based on the analysis of primary data collected by the book Author and selected secondary data related to digitalization and the development of employees' digital competencies, the evaluation of hypotheses will take place, which is one of the critical outcomes of the book.

1.5 Methodological Procedure of Research

The book's main objective was determined by defining the area of theoretical background and searching for appropriate literature that contributed to the solution to this objective. The methods of deduction and comparison were applied in the processing of the obtained data, which allowed going from the general to the specific and comparing the opinions of the authors to justify the opinion of the Author of the book on the issue under study. The theoretical background was defined through the analytical-synthetic cognitive procedure (i.e., complementary system analysis and synthesis).

An overview of the individual statistical methods used for further analysis and evaluation of the data obtained in the empirical research is presented in Sect. 1.7.

The research focused on digitalization and the related T&D of employee competencies. The literature review confirmed that digital competencies are fundamental and desirable in most jobs in different sectors of the economy during the COVID-19 pandemic, as both employers and employees agree (see Sects. 4.2.2, 5.1, and 5.2).

The research in this book was conducted from two perspectives: the employers (i.e., Section A) and the employees themselves (i.e., Section B). The whole methodological procedure of the book is divided into the following 14 phases, listed below in Table 1.3.

A. Digitalization and employee competencies from the employer's perspective

Table 1.2 Theoretical framework of the book

Key concepts	Author	Main outputs	Section
Digitalization	[2, 3, 5, 23, 25, 29, 32, 38, 43, 44, 47, 61, 66, 68–70, 72]	The term's definition in the literature, the historical development of digitalization due to the digital revolution, and the proposal of one's definition of digitalization to bridge the gap in existing knowledge	2.1, 2.2
Digital transformation	[8, 17, 19, 63, 71]	Definition of the term in scientific publications; emphasis on the differences in digital transformation from the concepts of digitization and digitalization	2.2
DT	[4, 7, 14, 15, 27, 30, 39, 40, 46, 53, 54, 57, 59, 62, 64]	Definition; classification of DTs; impact of DTs on business processes and activities of companies	2.3
Digital HRM	[6, 9, 15, 26, 39, 40, 41,48, 52, 58]	Definition and essence of the term; relation to the Fourth Industrial Revolution; distinctive features; areas of HRM affected by digitalization	3.1, 3.2
Competency model	[1, 13, 16, 34–37, 51, 55, 56, 60, 67]	Definition of the term; typology of competency models and their creation process, use in HRM, and competency models influenced by digitalization	4.2, 4.2.1, 4.2.3
Digital competence	[8, 10, 12, 18, 20, 22, 24, 42, 65]	Definition and essence of the term; role and meaning in the competency model; introduction of existing digital competency models	4.2.2, 4.2.4

Table 1.3 Methodological procedure for book preparation

	A. Digitalization and employee competencies from the employer's perspective
1	Development of the pilot research concept and selection of the target group
2	Modification of the research design and specification of the target group
3	Creating a custom enterprise database for research
4	Selecting sectors for deeper exploration
5	Choosing how to reach respondents
6	Seeking alternative ways to disseminate information about ongoing research
7	Implementation of quantitative research and evaluation of primary data
8	Developing the concept of in-depth interviews
9	Implementation of qualitative research and evaluation of primary data
	B. Digitalization and employee competencies from the employees' point of view
10	Research design and selection of target groups
11	Creating questionnaires and conducting quantitative research
12	Expanding the target group and conducting quantitative research
13	Analysis and evaluation of primary data
14	Formulating recommendations for employers

Phase 1: Design the pilot research and select the target group. The main objective of the pilot research was to find out how business entities in the CR view the use of DT and its impact on employees' competencies. The basic set of respondents consisted of persons over 18 years of age who were primarily employed as HR managers or owners of small and medium-sized enterprises (SMEs) or were self-employed and participated in the Knowledge Transfer Week at FE TUL from 16 to 20 September 2019. The structured anonymous questionnaire was prepared in paper form and included the following types of questions:

- Polytomous closed questions with one or more answer choices;
- A five-point Likert scale was used to measure respondents' attitudes about how much they agree or disagree with specific statements;
- Identification questions characterizing the enterprises involved in the research.

Phase 2: Revising the research design and specification of the target group. Concerning the comments received during the pilot research, the questionnaire was modified in March–April 2020, namely reformulating the existing questions, adding new questions (i.e., open-ended questions and semantic differential scales), and changing the strategy for selecting respondents. The book's Author expanded the basic sample to all active enterprises in the CR regardless of their size in terms of the number of employees and location sector of the national economy. The territorial scope of the survey was expanded because the pilot research, due to the location of FE TUL in the Liberec region of the CR, included mainly representatives of local enterprises, which would not allow the generalization of the results obtained within the country. The focus on enterprises of all sizes operating in all sectors of the economy was motivated by the need to determine whether there are differences between them in digitalization concerning these two criteria.

Phase 3: Creating a database of research companies. Data on businesses were extracted from the MagnusWeb database according to the criteria set out in Table 1.4. Access to the database was provided through a license purchased for all employees and Ph.D. students working at the DBA at FE TUL, where the Author of the book works.

The preparation of the surveyed enterprises' database consisted of eight stages. Phases 1, 2, 3, and 4 were carried out using the Filtering dialog box directly in the Magnus Web database. The filtered results were then downloaded in.xls format. Phases 5, 6, and 7 were already implemented using Microsoft (MS) Excel 2016 filters. Since the changes in the data of Czech companies are not immediately visible in the MagnusWeb database and the database still needs to be updated daily, the Author of the book still verified the functionality of the companies' websites. Subsequently, the book's Author also compared the contact e-mail addresses listed in the MagnusWeb database and on the companies' websites and updated them in an Excel file. Thus, the implementation of phase 8 took 5.5 weeks.

Phase 4: Selecting sectors for deeper exploration. This need was triggered by the size of the established enterprise database and based on the secondary data analysis carried out in Sect. 5.2. The sectors of the global economy experienced accelerated digitalization due to the COVID-19 pandemic (see Sect. 2.4). As a result, the base set has been adjusted for enterprises operating in the following sectors according to the CZ-NACE classification of economic activities:

- C—Manufacturing, in particular, subcategories involving the automotive industry:

Table 1.4 Procedure for preparing the prepared database of surveyed enterprises

Stages	Criterion	Number of enterprises
1	All legal and natural persons in the MagnusWeb database 1 June 2020	7,821,363
2	Active legal entities are identified in the MagnusWeb database	840,581
3	Active enterprises identified among legal persons	580,318
4	Records after removal of enterprises with no employees and enterprises for which the number of employees is not given	194,966
5	Records following the removal of undertakings in a state of bankruptcy or liquidation	192,954
6	Number of enterprises without significant adverse events in history	156,181
7	Records after removing businesses without a listed email address and website	93,910
8	Records after removing businesses that have non-functioning websites (i.e., the link to the website did not open, the website is blocked, the company has ceased operations)	27,780

- 22—Manufacture of rubber and plastic products;
- 25—Manufacture of fabricated metal products, except machinery and equipment;
- 28—Manufacture of machinery and equipment n.e.c;
- 29—Manufacture of motor vehicles (except motorcycles), trailers, and semi-trailers.
- F—Construction;
- K—Finance and Insurance.

The final stage's base consisted of 3270 construction companies, 2104 companies operating in the automotive industry, and 69 banks and insurance companies from all regions of the CR.

Phase 5: Choosing how to reach respondents. Given the size of the population and its geographical distribution, the CAWI method was chosen. The questionnaire was created using the survey software Survio. Its paid Business version allowed the collection of up to 10,000 completed questionnaires per month with subsequent export of responses in MS Excel 2016 in xls format. Under the bulk email distribution terms with TUL domains, the number of messages sent with an accompanying letter, a web link to the questionnaire, and a set deadline date was limited to 200 emails per day.

Phase 6: Seeking alternative ways to disseminate information about ongoing research. In order to reduce the risk of low return rates, several potential partners were approached to support data collection:

- The District Chamber of Commerce in Liberec (DCC in Liberec, CR)—an interest in the topic under study, placing a link to the questionnaire on the website and social networks.
- The District Chamber of Commerce in Jablonec nad Nisou (DCC in Jablonec nad Nisou, CR)—sending a questionnaire to all members, arranging contacts with HR managers in selected companies to conduct in-depth interviews.
- Association of Small and Medium-sized Enterprises (ASME) of the CR– the application was rejected because many applicants requested sending out various invitations, questionnaires, and other information that could not be physically provided.
- Czech Banking Association—the request remained unanswered.
- Czech Association of Insurance Companies—the request remained unanswered.
- Union of Building Entrepreneurs–the request remained unanswered.
- Sixty employees of Czech banks and insurance companies, 16 of them are graduates of TUL—a request sent to a personal LinkedIn profile went unanswered.

Phase 7: Implementation of quantitative research and evaluation of primary data. All enterprises identified in Phase 4 were subsequently approached to complete the questionnaire from July–October 2020. One hundred seventy-two enterprises (i.e., two insurance companies, 109 construction enterprises, and 61 automotive enterprises) had a non-functional email address and were subsequently excluded from the pool of respondents. As a result, data were obtained from 107 companies of all sizes from all regions of the CR. Selected data obtained from 27 construction companies, 39 automotive companies, and 10 banks and insurance companies were subject to further processing in MS Excel 2016. All statistical testing (i.e., Pearson's chi-square test of independence, Fisher's exact test, Mann–Whitney U test, and Spearman's correlation coefficient) was performed using IBM SPSS Statistics 25 statistical software.

Phase 8: Developing the concept of in-depth interviews. The semi-structured interview was chosen to give respondents more flexibility and freedom to express their views than in a structured interview and more organization and consistency than in an unstructured interview. That is, the structure of the in-depth interview was prepared in advance, and respondents were familiar with it but could have followed it more strictly. Based on the results of the literature search and secondary data analysis presented in Sects. 5.1 and 5.3, 17 open-ended questions were prepared, divided into five research areas, which are:

- Information about the company;
- General competence requirements for employees;
- Digitalization in HRM companies;
- Developing employees' digital competencies and the investments needed to do so;
- The impact of the COVID-19 pandemic on digitalization and innovation in society.

Phase 9: Implementation of qualitative research and evaluation of primary data. Four in-depth interviews with HRM experts were conducted in the selected companies between October and December 2020. These were one in-depth interview with three HR managers of a selected Czech bank and three interviews with HR experts in three automotive companies in the Liberec region of the CR. One face-to-face interview lasted 45 min. The interviewees' answers were recorded on a voice recorder with their verbal consent.

Subsequently, the obtained responses were converted into a word-by-word text format in MS Word and translated from Czech to English. During the transcription of the data, data anonymization was used to protect the names of the respondents and the companies they represented. Furthermore, a manual coding and analysis method (i.e., paper and pencil) was used. Finally, the collected data was sorted and compared to each other to find relationships and interpret them by the book Author. Based on the results obtained, the context of digitalization and employee competence development was further described in two case studies presented in Sects. 6.13 and 61.4.

B. Digitalization and employee competencies from the employee perspective

Phase 10: Research design and selection of target groups. Based on the theoretical findings and data from the research studies presented in Sect. 5.1, the research was decided to focus on two target groups. The selection of these two target groups was influenced by the fact that the Author of this book was a Ph.D. student at the DBA at FE TUL. The topicality of the topic under investigation and the fact that a survey on digital readiness had yet to be conducted among faculty staff or students at FE TUL before served to support the topic. The first target group consisted of employees in their primary employment among CS students studying at FE TUL in the academic year 2020/2021 who work in Czech companies. The second target group was employees working in the academic environment at FE TUL.

A population of 179 students of all years was specified based on data from the information system designed to administer the college's IS STAG student records on April 28, 2021. Eight male and female maternity or parental leave students were subsequently excluded from the respondent pool. Subsequently, the base sample was reduced to 171 respondents. Based on the data as of June 24, 2021, FE TUL had 90 internal and 10 external staff members. At the same time, there were 85 academic and 15 non-academic staff. Five external staff without a university email address were excluded from the research sample. Thus, the target group consisted of 80 academic staff.

Phase 11: Creation of questionnaires and implementation of quantitative research. Similar to the research on Czech companies, the method of electronic questioning was chosen here, and CAWI used the Survio survey software for both target groups. A random selection of respondents from among the employed CS students was implemented using the RANDBETWEEN function in MS Excel 2016. A structured questionnaire was prepared from March to April 2021. Subsequently, the questionnaire was distributed to randomly selected respondents between late April and the first half of June 2021 to ensure the results' representativeness. A total of 31 responses were obtained. For employees at the FE TUL, the survey was directed at the target respondents, resulting in 37 responses.

Phase 12: Expanding the focus group and conducting quantitative research. It was decided to supplement the research with primary data obtained abroad at the request of the rector of a German university, who also expressed interest in conducting a study to explore the topic in more depth. The name of the university was anonymized. The questionnaire used during the employee survey at FE TUL was first consulted with the university's Centre for E-Learning, the Rector, and the Vice-Rector for Education and International Relations and subsequently modified. The primary population of this research was 153 employees involved in teaching at a selected German university. The book's Author created the electronic questionnaire using the survey software available at this university and distributed it in German via the university. 79 complete and six partial responses to the questionnaire were obtained because the answers to all questions in the questionnaire were optional.

Phase 13: Analysis and evaluation of primary data. The processing and evaluation of the obtained data, including statistical testing of hypotheses, was carried out using MS Excel 2016 and IBM SPSS Statistics 25 statistical software.

Phase 14: Formulating recommendations for employers. Based on the responses to the surveys conducted and presented in Chap. 7, the Author of the book proposed recommendations for employers.

1.6 Sources and Methods of Data Collection

For this book, a combination of primary and secondary data is used. For clarity, Fig. 1.2 graphically presents the whole primary data collection process.

Secondary data, including Czech and foreign scientific articles, scientific literature, and research reports, were used to investigate the current state of the issue, prepare questionnaires, and create a database of business entities for primary data collection.

Primary data were collected using a written and electronic questionnaire survey and semi-structured in-depth interviews. The written questionnaire was used only for pilot research and was aimed at obtaining direct feedback from respondents for further possible modifications and additions. The electronic questionnaire survey was chosen as the main method of primary data collection for several reasons: its impartiality, time-saving nature, more accessible outreach to respondents, and obtaining responses in electronic form for subsequent immediate processing [28].

The semi-structured in-depth interview method was chosen to give respondents more space for their expression and leave the possibility for uncovering new aspects of the topic under study. For the same reason, semi-structured, in-depth interviews were conducted after the quantitative research to enrich and complement the findings. The choice of the case study method was motivated by the need to ensure a holistic approach, to obtain answers to questions such as "how" and "why", and to capture the research topic in the context of actual business practice [33].

The MagnusWeb business database and business websites were data sources for contacting businesses. Subsequent primary data collection was carried out in two main ways, presented in more detail in Sect. 1.5.

A. Primary data collection from the employers' perspective

The pilot research was prepared and implemented by the Author of the book in the framework of the participation in the TJ02000206 "Development of skills necessary for the digital transformation of business" research project of the Technology Agency of the CR. Respondents were selected randomly during the Knowledge Transfer Week, which took place at FE TUL in autumn 2019, and then asked to fill in a paper questionnaire. An improved questionnaire version was processed and used for data collection in Czech companies in 2020.

During the preparation of the questionnaire survey and interviews, cooperation was established with the directors of the DCC in Liberec and the DCC in Jablonec nad Nisou. They supported the distribution of the questionnaire among their members and recommended that HR managers in several companies willing to participate in in-depth interviews explore the topic further.

The collection of primary data in Czech enterprises started with preparing the database of business entities operating in the CR according to the criteria mentioned in Sect. 1.5. Then, it continued with the selection of specific sectors and the implementation of the survey. In order to protect the privacy of data, the survey did not collect any personal data of the respondents. Of the 107 responses, data obtained from representatives of 27 construction and 39 automotive companies as well as ten large (i.e., with more than 250 employees) and medium-sized (i.e., with 50–249 employees) banks and insurance companies operating in Prague (the CR) were selected for further processing in MS Excel 2016. In total, there were 76 respondents. Data obtained from representatives of three small banks and insurance companies (i.e., with 1–9 employees) were excluded from the analysis because these companies did not consider the development of their employees' competencies in connection with digitalization as an important and topical issue within their long-term HRM strategy. According to the Author of the book, reasons for the low return rate of the questionnaire (i.e., 2.0%) include:

- Questionnaire length (i.e., 19 questions);
- Implementation during the Covid-19 pandemic;
- The reluctance of respondents to answer these types of surveys;
- Respondent workload;
- The risk is that the email may have been included in spam.

However, the main reason is that the data collection occurred during the first six months of the COVID-19 pandemic, when Czech employers, in an unpredictable situation, devoted all their time and energy to adapting to the new working conditions. Another critical role was the need to increase the intensity and speed of digitalization, which may have pressured employers and made them unwilling to share their experience in this area. Summary information regarding the primary data collection in Czech companies is presented in Table 1.5.

The qualitative research in the form of in-depth interviews was aimed to complement the findings of the quantitative questionnaire survey. During the period of the research, the focus of the companies was on understanding how employees use DTs and what support is needed to develop their digital competencies during the COVID-19 pandemic, as well as the need to limit face-to-face contacts in response to the ongoing COVID-19 pandemic had a significant impact on companies' interest in collaboration. As a result, the number of interviews conducted was limited to four. At the request of the respondents, the names of the companies interviewed have been anonymized.

B. Collection of primary data from the employees' point of view

Table 1.6 presents the main data on the surveys conducted to collect primary data from employees in the CR and abroad.

The survey of employees of Czech companies among students of all years of CS at FE TUL was conducted in the spring of 2021. From the end of April to the first half of June 2021, the link to the electronic anonymous questionnaire was sent to randomly selected respondents twice at the university e-mail address.

Primary data collection from staff in the academic environment began with research at FE TUL in the summer of 2021. As of 1 June 2021, there were 100 employees at FE TUL, of which 15 were non-academic staff. Given that the participation of non-academic staff was minimal (only three completed questionnaires were received), their responses were excluded from this research. At the same time, five

Table 1.5 Sources of primary data from the employer's perspective

Type of survey	Data type and method of collection	Date of data collection	Research sample	Sector according to CZ-NACE classification	Return rate of questionnaires
The pilot survey	A written, structured questionnaire, quantitative data	16–20 September 2019, FE TUL	48 self-employed entrepreneurs or employees were among the participants	In all sectors, respondents were randomly selected	62.5% (30 replies)
Research on Czech enterprises	An electronic structured questionnaire, quantitative data	8 July–19 October 2020	3,270 companies	F–Construction	2.0% (89 replies)
			2,104 companies	C–Manufacturing (codes 22, 25, 28, 29)	
			160 members of the DCC in Jablonec nad Nisou	All sectors	5.0% (8 replies)
			Members of the DCC in Liberec	All sectors	10 replies
			69 companies	K Finance and Insurance	18.9% (13 replies)
	Semi-structured in-depth interview, qualitative data	19 October, 17 December 2020	Three HR managers from three companies in the Liberec Region of the CR	C–Manufacturing (codes 22, 25, 28, 29)	X
		24 November 2020	Three HR specialists from one bank	K–Finance and Insurance	X

academic staff members were not invited to participate in the research as they were external employees and needed an e-mail address on the tul.cz domain.

It was followed by data collection abroad at the end of April 2022 from 153 selected employees of a German university. A new questionnaire was developed for this purpose. The questionnaire used for the research among academic staff at the FE TUL was taken as a basis. This questionnaire was first translated into English, and its content was consulted by the collaborators of the *Zentrum für eLearning* (angl. German Centre for E-Learning). It is a focal point for using DTs that offer comprehensive services to support teaching and learning and implement administrative processes at a selected German university. Subsequently, the questionnaire was sent to the Vice-Chancellor for Education for review, to International Affairs,

Table 1.6 Sources of primary data from an employee perspective

Type of survey	Data type and method of collection	Date of data collection	Research sample	Return rate of the questionnaire
Employee research (CS students at FE TUL)	An electronic structured questionnaire, quantitative data	28 April–10 June 2021	171 CS students at FE TUL; respondents were randomly selected	34.45% (31 replies)
Research of university staff	An electronic structured questionnaire, quantitative data	24 June–22 July 2021	80 academics at FE TUL	46.25% (37 replies)
	An electronic structured questionnaire, quantitative data	28 April–15 May 2022	153 academics were selected at a German university	55.6% (79 complete replies and six partial replies)

and to the Rector to adapt it to the different conditions of the academic environment in Germany. After incorporating the recommendations and comments received, the book Author translated the questionnaire into German using the online translator DeepL Translate and created the online questionnaire using the software Survey. It was the software available at the selected German university. The subsequent preparation for the electronic survey took six weeks. The detailed results were also forwarded to the Rector of the German university in the form of a research report in German and presented in English by the book Author to the Vice-Chancellor for Education and International Affairs. At the request of the Rector, the university's name was anonymized. The results were translated from German into English for further use in this book. Considering that all questions in this questionnaire were optional, a different number of responses were obtained for each question, specified in more detail when analyzing the primary data collected in Sect. 6.2.2.

1.7 Quantitative Methods Used in the Evaluation of the Collected Data

In the empirical part of the book, several quantitative methods were applied to analyze and subsequently evaluate the primary data obtained. Descriptive statistics methods were used for analysis. More complex quantitative and statistical methods in the form of parametric and non-parametric tests were chosen for further evaluation of the obtained primary data. However, parametric tests have more statistical power than non-parametric tests. Non-parametric tests complemented parametric tests because non-parametric tests do not require assumptions about the nature of the normal distribution of the random variables under study and are not sensitive to outliers [11]. The methods applied include Pearson's chi-square test of independence, Fisher's exact test, Mann–Whitney U test, rank correlation independence test, simple analysis of variance, Kruskal–Wallis H test, and Welch's test. The individual methods are described in more detail below.

Pearson's Chi-Square Test of Independence

Pearson's chi-square test of independence (i.e., Pearson's x^2 test of independence) is the most widely used in contingency tables when it is necessary to determine whether two random variables, X and Y (i.e., unpaired categorical variables) are independent. This test compares the observed and expected (i.e., theoretical) frequencies of each combination of random variables X and Y. In doing so, the expected frequencies are calculated assuming the null hypothesis H_0. The null hypothesis H_0 asserts that the random variables X and Y are independent. The alternative hypothesis H_1 states that there is a statistically significant relationship between the random variables X and Y. It can be written as follows:

H_0: Variables are independent.

H_1: Variables are dependent.

The magnitude of the difference between the expected and observed frequencies in any contingency table is determined using a test criterion x^2 for the Pearson's chi-square test of independence, where n_i are the empirical frequencies and $n\pi_{0,i}$ are the theoretical frequencies (see Eq. 1.1):

$$x^2 = \sum_{i=1}^{k} \frac{\left(n_i - n\pi_{0,i}\right)^2}{n\pi_{0,i}} \tag{1.1}$$

Subsequently, based on the probability distribution x^2, the statistical test's achieved significance level, also called the P-value, is calculated. In the next step, this is compared with the chosen significance level at which (α) are tested. If the P-value is less than or equal to α, H_0 is rejected. The number of degrees of freedom is calculated as the number of rows in the contingency table scaled by one multiplied by the number of columns in the contingency table scaled by one.

The test result can also be determined if the value x^2 calculated according to Eq. 1.1 is compared with the critical value of the distribution in chi-square tables at the significance level $\alpha = 0.05$ for the observed number of degrees of freedom. Suppose it turns out that the resulting value of the statistic x^2 exceeds the critical value, the null hypothesis is rejected [21].

The assumptions of Pearson's chi-square test of independence that need to be verified before using it are as follows:

- One cell of the contingency table must contain only one element of the sample, thus ensuring the independence of the observations listed in the contingency table;
- At least 80% of the expected frequencies must be greater than 5;
- 100% of the expected frequencies must be greater than 2 (for a four-field contingency table, i.e., of the 2 × 2 type, they should be greater than five, and it

is recommended not to fall below ten); these are called good approximation conditions;
- If the expected frequencies are less than 5, Fisher's exact test, which is not based on asymptotic methods, can be used [11, 49].

Failure to meet Pearson's chi-square independence test assumptions can lead to nonsensical conclusions.

The results of statistical hypothesis testing using Pearson's chi-square test of independence are presented in Sects. 6.1.1, 6.1.5 and 6.2.3.

Fisher's Exact Test

Fisher's exact test is designed primarily for four-field tables and for analyzing small data sets regardless of the data distribution. This test is based on calculating the exact (i.e., precise) probability that the same or more extreme result would have been obtained if the sum of the columns and rows in the contingency table had been kept. It is used when the excellent approximation conditions still need to be met. The null hypothesis H_0 states that the observed variables X and Y are independent, and in the case of its validity, the observed frequencies should match the expected frequencies. In contrast, the alternative hypothesis H_1 states that the observed variables X and Y are dependent.

The calculation of the test statistic then starts with calculating the probability ρ with which the resulting four-field table would be equally or more distant from hypothesis H_0 if the observed marginal frequencies were retained. The conservation of marginal frequencies can correspond to the same frequencies for each variant of the random variables X and Y. Probability ρ can be found using Eq. 1.2:

$$\rho = \frac{\binom{a+c}{a}\binom{b+d}{b}}{\binom{n}{a+b}} = \frac{(a+b)!(a+c)!(c+d)!(b+d)!}{n!a!b!c!d!} \tag{1.2}$$

The test statistic is determined using the following procedure. First, the probability ρ is calculated for all possible tables obtained at marginal frequencies. Subsequently, all probabilities ρ less than or equal to the P-value is summed. It is a situation in which the P-value assumes the role of the resulting test statistic, which is compared to the chosen significance level of $\alpha = 0.05$. If the P-value turns out to be less than the null hypothesis of independence of the randomly selected variables X and Y are rejected.

Sections 6.1.1, 6.1.5, and 6.2.3 present the results of statistical hypothesis testing using Fisher's exact test.

Mann–Whitney U Test

The Mann–Whitney U test is a non-parametric test based on the U test criterion that detects the number of inversions. It is also called the test of the agreement of means or medians. First, two independent random sets $(x_1, x_2, \ldots x_{n1})$ and $(y_1, y_2, \ldots y_{n2})$ are selected from continuous distributions with the same variance and shape. The condition is that $n_1 \geq n_2$. The null hypothesis H_0 states that the medians of the two samples coincide. The alternative hypothesis H_1 states that the two random samples do not have identical medians.

When calculating the test criterion, it is first necessary to arrange all the values $n_1 + n_2$ from both files in ascending order and arrange them so that the lowest value is assigned the value one and the highest value is assigned the value $n_1 + n_2$. The sum of the order of the values $x_1, x_2, \ldots x_{n1}$ is denoted as R_1, and the sum of the order of the values $(y_1, y_2, \ldots y_{n2})$ is denoted as R_2. In doing so, the following Eq. 1.3 holds:

$$R_1 + R_2 = n(n+1)/2 \tag{1.3}$$

The calculation of the test statistic U is performed according to the following Eq. 1.4:

$$U_1 = R_1 - n_1(n_1+1)/2, U_2 = R_2 - n_2(n_2+1)/2 \tag{1.4}$$

It is also true that (see Eq. 1.5):

$$U_1 + U_2 = n_1 n_2 \tag{1.5}$$

The next step is to determine the test criterion (see Eq. 1.6), whose critical value of the distribution under the condition of the validity of the null hypothesis H_0 is obtained from the statistical tables for the significance level $\alpha = 0.05$:

$$R(X, Y) = \min(U_1, U_2) \tag{1.6}$$

If the test criterion value is equal to or lower than the relevant critical value from the statistical tables for the Mann–Whitney U test, the null hypothesis is rejected [31, 45].

Sect. 6.1.1 and 6.1.5 present the results of statistical testing of the research assumptions using the Mann–Whitney U test.

Test for Independence of Ordinal Correlation

The test for independence of ordinal correlation tests the value of Spearman's coefficient in the base set, which is referred to as ρ_s. The null hypothesis H_0 states that $\rho_S = 0$ and the dependence between X and Y is not monotonic. The alternative hypothesis H_1 states that $\rho_s < 0$ or $\rho_s > 0$. For the sample size $n < 10$ the extreme points of the critical region must be looked at in particular tables. For $n \geq 10$ test criterion can be used (see Eq. 1.7), which has a distribution t s $n - 2$ degrees of freedom:

$$t = \frac{r_s}{\sqrt{1 - r_s}} \sqrt{n-2} \tag{1.7}$$

At the significance level α the critical region is defined by the inequality $|t| > t_{1-\alpha/2}(n-2)$ or by a test criterion (see Eq. 1.8) that has under the null hypothesis H_0 a distribution F with 1 and $n-2$ degrees of freedom:

$$F = \frac{r^2}{1-r^2}(n-2) \tag{1.8}$$

At the significance level α the critical region is defined by the inequality $F = F_{1-\alpha}(1, n-2)$ [50].

The sample Spearman's rank-order correlation coefficient measures the strength of the dependence of two ordinal signs or an ordinal and metric signs X_i and Y_i selected randomly from a bivariate distribution. Each can take on its order of magnitude, denoted as R_X and R_Y. If with increasing values of X_i there is an increase in the values Y_i, the orders of the two quantities coincide, i.e. $R_{Xi} = R_{Yi}$. If Y values decrease in this situation, it is clear that the orders of the two quantities are opposite. If the quantities X_i and Y_i are independent, their values are arranged entirely randomly. The consequence is that the Spearman's correlation coefficient, denoted by the letter r_s (r from ranking), is found using ranking differences ($R_{Xi} - R_{Yi}$) denoted as d_i.

Equitation 1.9 gives the method of its calculation:

$$r_s = 1 - \frac{6\sum d_i^{\ 2}}{n\left(n^2-1\right)} \tag{1.9}$$

In Eq. 1.9 d_i represents the difference of the measured values X_i and Y_i and n is the order number.

Spearman's coefficient of correlation of ordinal numbers can take values in the interval from -1 to $+1$. If the order matches, r_s reaches its maximum limit and equals 1. In a situation where the order is reversed, r_s it reaches its minimum limit and is equal to -1. If the r_s equals 0, it can be argued that there is no dependence between the observed quantities X and Y and their orders are randomly interspersed [31].

The results of statistical testing of the research assumptions using Spearman's correlation coefficient are presented in Sect. 6.1.1.

Anova

ANOVA is a statistical method generalizing the T-test for two independent samples. ANOVA is used to test the hypothesis that the means of a given numerical variable are the same in more than two independent groups. These groups are defined based on a U-factor (i.e., K categories of one categorical variable) and may vary in size. Subsequently, a simple analysis of variance aims to identify the factors that significantly affect the variable of interest. The nxp data matrix is decomposed vertically into groups, the number of which depends on the number of categories of one factor or the number of combinations of all factors that were used [45].

The null hypothesis H_0 states that the means (medians) of the groups are equal. The alternative hypothesis H_1 states that some two groups' averages (means) are not equal. It can be written as follows:

H_0: Group averages are the same.

H_1: The averages in some two groups are different.

The test criterion (i.e., the proportion of the sum of squares representing the between-group Q_B and intra-group Q_E variability of the explained variable is the basis for testing the hypothesis of independence of a single measurable variable from a single factor U of K categories (see Eq. 1.10):

$$F = \frac{Q_B/(K-1)}{Q_B/(n-K)} \tag{1.10}$$

In order to apply ANOVA, it is first necessary to verify that the following conditions are met:

- Condition 1: The dependent variable should be at least ordinal;
- Condition 2: The independent variable or factor should consist of two or more categorical independent groups;
- Condition 3: The independence of observations should be ensured, which means that between observations within groups or between groups;
- Condition 4: There should be no significant outliers;
- Condition 5: Individual selections should come from a normal distribution;
- Condition 6: The same variance of individual selections should be maintained in the groups (i.e., homoskedasticity).

Conditions 4, 5, and 6 can be verified using IBM SPSS Statistics 25. The ordinal scales should have no outliers (i.e., Condition 4).

To verify Condition 5, a histogram and the Shapiro–Wilk normality test can be used for visual inspection. A simple analysis of variance is quite "robust" to violations of normality, which means that this assumption may be violated somewhat if the size of the groups is equal or nearly so and still gives valid results. When normality is significantly violated, using the Kruskal–Wallis H test, also called nonparametric ANOVA, is appropriate. The Kruskal–Wallis H test also extends the Mann–Whitney U test.

Condition 6 can be verified using Levene's test of the agreement of variances in groups. If condition 6 is not satisfied, alternative tests, such as the Welch's test, which is part of the IBM SPSS Statistics 25 statistical program, and homoskedasticity, are available.

Kruskal–Wallis H Test

As [45, p. 146] states:

> "The Kruskal-Wallis H test for independent samples is based on expressing the between-group variability of the rankings assigned to the original observations before they are sorted into groups".

The null hypothesis H_0 states that the medians of all groups are the same. The alternative hypothesis H_1 states that some two groups' medians (i.e., means) are not equal. It can be written as follows:

H_0: Medians in the groups are the same.

H_1: Medians in some two groups are different.

The Kruskal–Wallis H test also applies to ordinal variables by ordering all observed values by magnitude. The test criterion is as follows (see Eq. 1.11):

$$H_{KW} = \frac{12}{\mathrm{n(n+1)}} \sum_{i}^{r} \frac{R_i}{n_i} - 3(n+1) \tag{1.11}$$

In Eq. 1.9, $i = 1, 2, \ldots, r$ denotes the sum of the ordinal numbers in each sample R_i. The Kruskal–Wallis H test is appropriate to use only if the data satisfy the four assumptions necessary to provide a valid result:

- The dependent variable should be at least ordinal;
- The independent variable or factor should consist of two or more categorical independent groups;
- Independence of observations should be ensured, meaning that there is no relationship between observations within groups or between groups;
- It is necessary to determine whether the distributions in each group (i.e., the distribution of scores for each group of the independent variable) have the same shape (which also implies the same variability).

Sections 6.1.5 and 6.2.3 present the results of statistical testing of the research hypotheses using ANOVA.

References

1. Armstrong, M., Taylor, S.: Armstrong's handbook of human resource management practice. Kogan Page, London, New York (2020)
2. Balcerzak, P., Pietrzak, B.: Digital economy in Visegrad countries: Multiple criteria decision analysis at regional level in the years 2012 and 2015. J. Com. **9**, 5–18 (2017). https://doi.org/10.7441/joc.2017.02.01
3. Balsmeier, B., Woerter, M.: Is this time different? How digitalization influences job creation and destruction. Res. Pol. **48**, 1–10 (2019). https://doi.org/10.1016/j.respol.2019.03.010
4. Baptista, J., Stein, M., Klein, S.: Digital work and organisational transformation: Emergent digital/human work configurations in modern organisations. J. Strat. Inf. Syst. **29**, 101618 (2020). https://doi.org/10.1016/j.jsis.2020.101618
5. Baraldi, E., Nadin, G.: The challenges in digitizing business relationships. The construction of an IT infrastructure for a textile-related business network. Technology. **26**, 1111–1126 (2006). https://doi.org/10.1016/j.technovation.2005.09.016
6. Berber, N., Đorđević, B., Milanović, S.: Electronic human resource management (e-HRM). Strat. Man. **23**, 22–32 (2018)
7. Bharadwaj, A., El Sawy, O., Pavlou, P.: Digital business strategy: Towards a next generation of insights. Mis Quart. **37**, 471–482 (2013). https://doi.org/10.25300/MISQ/2013/37:2.3

8. Bikse, V., Lusena-Ezera, I., Rivza, P.: The development of the digital transformation and relevant competencies for employees in the context of the impact of the Covid-19 pandemic in Latvia. Sustainability. **13**, 9233 (2021). https://doi.org/10.3390/su13169233
9. Bissola, R., Imperatori, B.: HRM 4.0: The digital transformation of the HR department. In: Cantoni, F., Mangia, G. (eds.) Human resource management and digitalization, pp. 51–72. Routledge, Abingdon, New York, Torino (2019)
10. Bondar, I., Komarnitskyi, I., Rusavska, V.: Formation and development of digital skills in university education. Lapl. Em. Rev. **7**, 47–56 (2021). https://doi.org/10.24115/S2446-622020217Extra-B882p.47-56
11. Borůvková, J., Horáčková, P., Hanáček, M.: Statistika v SPSS (Statistics in SPSS). Polytechnic University, Jihlava (2014)
12. Calvani, A., Cartelli, A., Fini, A., Ranieri, M.: Models and instruments for assessing digital competence at school. J. E-Learn. Knowl. Soc. **4**, 186–193 (2008). https://doi.org/10.20368/1971-8829/288
13. Cardy, R., Selvarajan, T.: Competencies: Alternative frameworks for competitive advantage. Bus. Hor. **49**, 235–245 (2006). https://doi.org/10.1016/j.bushor.2005.09.004
14. Casseta, E., Monarca, U., Dileo, I., Di Berardino, C., Pini, M.: The relationship between digital technologies and internationalisation: Evidence from Italian SMEs. Ind. Inn. **27**, 311–339 (2020). https://doi.org/10.1080/13662716.2019.1696182
15. Chytiri, A.P.: Human resource managers' role in the digital era. Spoud. J. Ec. Bus. **69**, 62–72 (2019)
16. Clempner, J., Gutierrez, A.: Prototyping for selecting people: A competencies model. In: Hamza M.H. (eds.) Proceedings of the 23rd international conference on modelling, identification, and control, pp. 485–494. Acta Press, Calgary (2004)
17. Fenech, R., Baguant, P., Ivanov, D.: The changing role of human resource management in an era of digital transformation. J. Man. Inf. Dec. Sci. **22**, 166–175 (2019)
18. Ferrari, A.: Digital competence in practice: An analysis of frameworks. Publication office of the EU. https://doi.org/10.2791/82116. (2012). Accessed 30 June 2024
19. Gong, C., Ribiere, V.: Developing a unified definition of digital transformation. Techn. **102**, 102217 (2021). https://doi.org/10.1016/j.technovation.2020.102217
20. Grevtseva, G., Bashtanar, I., Isakova, T., Balikaeva, M., Kirillova, Y.: The digital information competence of an individual: Historical aspect. In: Shaposhnikov, S. (ed.) Proceedings of the international conference on quality management, transport and information security, information technologies, pp. 474–476. Institute of Electrical and Electronic Engineers, Piscataway (2020)
21. Hindls, R., Hronová, S., Seger, J.: Statistika pro ekonomy (Statistika pro economy). Professional Publishing, Prague (2003)
22. Ilomäki, L., Paavola, S., Lakkala, M., Kantosalo, A.: Digital competence — an emergent boundary concept for policy and educational research. Ed. Inf. Techn. **21**, 655–679 (2016). https://doi.org/10.1007/s10639-014-9346-4
23. Ivanová, E., Žárská, V., Masárová, J.: Digitalization and human capital development. Entr. Sust. Iss. **9**, 402–415 (2021). https://doi.org/10.9770/jesi.2021.9.2(26)
24. Janssen, J., Stoyanov, S., Ferrari, A., Punie, Y., Pannekeet, K., Spoel, P.: Experts' views on digital competence: Commonalities and Differences. Comp. Ed. **68**, 473–481 (2013). https://doi.org/10.1016/j.compedu.2013.06.008
25. Kagermann, H.: Change through digitization — value creation in the age of Industry 4.0. In: Albach, H., Meffert, H., Pinkwart, A., Reichwald, R. (eds.) Management of permanent change, pp. 23–45. Springer Gabler, Wiesbaden (2015)
26. Karaboğa, T., Karaboğa, H., Basar, D., Zehir, S.: The digital transformation journey of HR: The effect of big data and artificial intelligence in HR strategies and roles. In: Pejic-Bach, M., Dogru, C. (eds.) Management strategies for sustainability, new knowledge innovation, and personalized products and services, pp. 94–115. IGI Global, Hershey (2022)
27. Khin, S., Ho, T.: Digital technology, digital capability and organizational performance: A mediating role of digital innovation. Int. J. Inn. Sci. **11**, 177–195 (2019). https://doi.org/10.1108/IJIS-08-2018-0083

28. Kozel, R., Mynářová, L., Svobodová, H.: Moderní metody a techniky marketingového výzkumu (Modern methods and techniques of marketing research). Grada Publishing, Prague (2011)
29. Legner, S., Eymann, T., Hess, T.: Digitalization: Opportunity and challenge for the business and information systems engineering community. Bus. Inf. Syst. Eng. **59**, 301–308 (2017). https://doi.org/10.1007/s12599-017-0484-2
30. Li, C., Shing Ip, H.: Defining virtual reality enabled learning. Int. J. Inn. Learn. **31**, 291–306 (2022). https://doi.org/10.1504/IJIL.2022.122128
31. Litschmannová, M.: Úvod do statistiky (Introduction to statistics). Technical University, Ostrava (2011)
32. Mancini, D., Lamboglia, R., Castellano, N.G., Corsi, K.: Trends of digital innovation applied to accounting information and management control systems. In: Corsi, K., Castellano, N.G., Lamboglia, R., Mancini, D. (eds.) Reshaping accounting and management control systems, pp. 1–19. Springer, Cham (2017)
33. Mareš, J.: Tvorba případových studií pro výzkumné účely (Creating case studies for research purposes). Ped. **65**, 113–142 (2015)
34. Markus, L., Cooper-Thomas, H., Allpress, K.: Confounded by competencies? An evaluation of the evolution and use of competency models. New Zel. J. Psych. **34**, 117–126 (2005)
35. Marrelli, A., Tondora, J., Hoge, M.: Strategies for developing competency models. Adm. Pol. In Ment. Health **32**, 533–561 (2005). https://doi.org/10.1007/s10488-005-3264-0
36. Matuska, E.: Human resource management system based on competences — Scania case study. Hum. Res. Man. Erg. **6**, 130–144 (2012)
37. McLagan, P.: Competency models. Train. Dev. J. **34**, 22–26 (1980)
38. Milošević, N., Dobrota, M., Rakočević, R.: Digital economy in Europe: Evaluation of countries' performances. Proc. Fac. Ec. Rej. **36**, 861–880 (2018). https://doi.org/10.18045/zbefri.2018.2.861
39. Molotkova, N., Makeeva, M., Khazanova, D.: Digitalized personnel management. In: Mantulenko, V. (ed.) The European proceedings of social and behavioural sciences, pp. 757–767. Future Academy, Samara (2019). https://doi.org/10.15405/epsbs.2019.03.75
40. Moorthy, J., Lahiri, R., Biswas, N., Sanyal, D., Ranjan, J., Nananth, K., Ghosh, P.: Big data: Prospects and challenges. Vikalpa. **40**, 74–96 (2015). 10.1177%2F0256090915575450
41. Negi, P., Kaur, J.: The digital workplace: Implications for human resource management. Int. J. Man. Techn. Eng. **9**, 1491–1500 (2019)
42. Oberländer, M., Beinicke, A., Bipp, T.: Digital competencies: A review of the literature and applications in the workplace. Comp. Ed. **146**, 103752 (2020). https://doi.org/10.1016/j.compedu.2019.103752
43. Ocampo, S.: Digitalisation and technological convergence from the sociological perspective of general social systems theory. Rev. Gest. De las Pers. Tech. **5**, 153–165 (2012)
44. Parviainen, P., Tihinen, M., Käriäinen, J.: Tackling the digitalization challenge: How to benefit from digitalization in practice. Int. J. Inf. Syst. Proj. Man. **5**, 63–77 (2017). https://doi.org/10.12821/ijispm050104
45. Pecáková, I.: Statistika v terénních průzkumech (Statistics in field surveys). Professional Publishing, Prague (2008)
46. Pousttchi, K., Gleiss, A., Buzzi, B., Kohlhagen, M.: Technology impact types for digital transformation. In: Becker, J., Novikov, D. (eds.) Proceedings of the 21st conference on business informatics CBI 2019, vol. 2, pp. 487–494. Institute of Electrical and Electronics Engineers, Piscataway (2019)
47. Rachinger, M., Rauter, R., Müller, C.: Digitalization and its influence on business model innovation. J. Man. Techn. Man. **30**, 1143–1160 (2019). https://doi.org/10.1108/JMTM-01-2018-0020
48. Rana, T., Bhakar, S., Kaurav, R., Agraval, V.: Digitalization of human resource practices—an emerging trend. In: Bhakar, S.S., Kaurav, R.P.S., Agraval, V., Gupta, M. (eds.) Proceedings of the 10th international conference of digital strategies for organizational success, pp. 1173–1183. (2019). https://doi.org/10.2139/ssrn.3323740

49. Reiterová, E.: Statistika pro nelékařské zdravotnické obory (Statistics for non-medical health disciplines). Palacky University, Olomouc (2016)
50. Řezanková, H.: Analýza dat z dotazníkových šetšení (Analysis of data from questionnaire surveys). Professional Publishing, Prague (2010)
51. Sampson, D., Fytros, D.: Competence models in technology — enhanced competence-based learning. In: Adelsberger, H., Kinshuk, H., Pawlowski, J., Sampson, D. (eds.) Handbook on information technologies for education and training. International handbook on information systems, pp. 155–177. Springer, Berlin, Heidelberg (2008). https://doi.org/10.1007/978-3-540-74155-8_9
52. Sergi, B., Klučnikov, A., Popkova, E.: Creative abilities and digital competencies to transitioning to Business 4.0. J. Bus. Res. **153**, 401–411 (2022). https://doi.org/10.1016/j.jbusres.2022.08.026
53. Shi, M., Zhang, C., Hu, J.: An innovative marketing strategy: Adopting AR VR in jewelry sales. In: Patnaik, S. (ed.) Proceedings of the 2nd international conference on applied machine learning ICAML 2020, pp. 333–338. IEEE Computer Society, Los Alamitos (2020)
54. Shpak, N., Maznyk, L., Dvulit, Z., Seliuchenko, N., Dragan, O., Doroshkevych, K.: Influence of digital technologies on the labor market of HR specialists. In: Vysotska, Y., Kupriianov, O., Kanishcheva, O., Cherednichenko, O., Hamon, T., Gradar, N. (eds.) Proceedings of the 6th international conference on computational linguistics and intelligent systems COLINS 2022, pp. 1475–1487. CEUR-WS, Gliwice (2022)
55. Singh Chouhan, V., Srivastava, S.: Understanding competencies and competency modelling—a literature survey. IOSR J. Bus. Man. **16**, 14–22 (2014)
56. Skorková, Z.: Competency models in public sector. Proc. Soc. Beh. Sci. **230**, 226–234 (2016). https://doi.org/10.1016/j.sbspro.2016.09.029
57. Spremic, M.: Governing digital technology—how mature IT governance can help in digital transformation?. Int. J. Ec. Man. Syst. **2**, 214–223 (2017)
58. Strohmeier, S.: Digital human resource management: A conceptual clarification. Germ. J. Hum. Res. Man. **34**, 345–365 (2020). https://doi.org/10.1177/2397002220921131
59. Šimec, A., Ibrahimov, M., Aleksic, A., Dukic, D.: Cyber-attacks and internet of things as a threat to critical infrastructure. In: Šimec, A., Ibrahimov, M., Aleksic, A., Dukic, D. (eds.) Proceedings of the 37th international scientific conference on economic and social development, pp. 108–111. Vazardin Development and Entrepreneurship Agency, Baku (2019)
60. Teodorescu, T.: Competence versus competency. What is the difference?. Perf. Impr. **45**, 27–30 (2006). https://doi.org/10.1002/pfi.4930451027
61. Tilson, D., Lyytinen, K., Sørensen, C.: Digital infrastructures: The missing is research agenda. Inf. Syst. Res. **21**, 748–759 (2010). https://doi.org/10.1287/isre.1100.0318
62. Tsou, H., Chen, J.: How does digital technology usage benefit firm performance? Digital transformation strategy and organisational innovation as mediators. Techn. An. Strat. Man. (2021). https://doi.org/10.1080/09537325.2021.1991575
63. Verhoef, P., Broekhuizen, T., Bart, Y.: Digital transformation: A multidisciplinary reflection and research agenda. J. Bus. Res. **122**, 879–901 (2021). https://doi.org/10.1016/j.jbusres.2019.09.022
64. Vial, G.: Understanding digital transformation: A review and a research agenda. J. Strat. Inf. Syst. **28**, 118–144 (2019). https://doi.org/10.1016/j.jsis.2019.01.003
65. Vuorikari, R., Kluzer, S., Punie, Y.: DigComp 2.2. The digital competence framework for citizens with new examples of knowledge, skills and attitudes. Publication Office of the EU. https://doi.org/10.2760/490274. (2012). Accessed 10 July 2024
66. Wachal, R.: Humanities and computers: A personal view. North Amer. Rev. **256**, 30–33 (1971)
67. Wang, Y., Wei, J.: A new framework of core competence of corporations: Hierarchy system model. In: Xiaofeng, J. (ed.) Proceedings of the 99th international conference on management science & engineering, pp. 341–345. Yichang (1999)
68. Westerlund, M.: Digitalization, internationalization and scaling of online SMEs. Techn. Inn. Man. Rev. 10, 48–57 (2020). https://doi.org/10.22215/timreview/1346

69. Yoo, H., Mody, B.: Predictions of downsizing in the U.S. local telephone industry. The Inf. Soc. **16**, 23–33 (2000)
70. Young, J.: Benefits of digitization and the impact of computer philosophies: Reaping the rewards of information. Int. J. Com. Appl. Techn. **1**, 10–14 (1988). https://doi.org/10.1504/IJCAT.1988.062642
71. Yucel, S.: Estimating the benefits, drawbacks and risk of digital transformation strategy. In: Guerro, G.E. (ed.) Proceedings of IEEE international conference on computational science and computational intelligence 2018, pp. 233–238. Institute of Electrical and Electronics Engineers, Piscataway (2018)
72. Zhongjun, L., Lijuan, H., Qihai, Z.: The reconstruction, exploration and innovation of logos of e-commerce. In: Yu, F., Luo, Q., Chen, Y., Chen, Z. (eds.) Proceedings of the international symposium on electronic commerce and security ISECS 2008, pp. 792–795. IEEE computer society, Los Alamitos (2008)

Chapter 2
Theoretical Foundations of Digitalization in the Context of Human Resources

Abstract This chapter presents the primary outcomes of the literature review on the existing knowledge concerning digitalization. Within each section, the concept of digitalization, which has developed over the decades, its definition, and application through the use of digital technologies in the daily activities of companies with respect to the changing working conditions and environment in response to the COVID-19 pandemic, are introduced.

2.1 Historical Development of the Concept of Digitalization

The concept of "digitalization" was born in the 1940s, when the digital revolution began. The digital revolution refers to any DT that can increase connectivity and collaboration between people and information systems without being limited to one sector [23]. Several key historical events preceded the birth of the digital revolution. The first was the development of the first electronic numerical integrator and computer (ENIAC) by Mauchly and Eckert of the University of Pennsylvania during World War II for the United States (US) Army. It was initially designed to integrate external ballistics equations step-by-step. However, because of its flexible control, it could be applied to many other types of computation simultaneously [26]. As a result of improvements in the ENIAC's technical parameters and the use of transistors, the inventors could move from public procurement to mass universal automatic computer (UNIVAC) production. The first private company to use UNIVAC in business, especially for payroll processing and production control, was General Electric's Kentucky, U.S., subsidiary in 1954 [15].

The second significant event that contributed to the development of the digitalization concept was the publication of the critical work of the mathematician and engineer [58] entitled "A Mathematical Theory of Communication". In his work, the American mathematician and computer scientist demonstrated that information can be reliably delivered through communication channels such as telephone lines and wireless connections. These ground breaking innovations provided the tools that

A. Mazurchenko, *The Impact of the Digitalization on the Human Resource Management*, Information Systems Engineering and Management 25,
https://doi.org/10.1007/978-3-031-76790-6_2

ushered in the Information Age, making Claude Shannon father of modern digital communication and information theory.

The next revolutionary step was the introduction of a fundamental principle that has governed the development of the microchip to date and is known in literature as Moore's Law [45]. An American chemist, engineer, and entrepreneur noticed that microchip transistors rapidly increase, exponentially increasing computing power while reducing chip cost. Moore's Law states that computing power doubles once every year. Moore's Law has further driven technological innovation and increased the scope and reach of digitalization.

The consequence of these fundamental changes was the drastic development of digitalization in the 1970s. During this period, the eventual transition of industrial control systems from conventional continuous control to computer discrete control occurred, thus beginning the gradual automation of production through the computerization of industrial operations [65].

The current concept of digitalization in its economic sense is based on the findings of [74] published in "The North American Review". In his work, this computer programming teacher at the American University of Iowa first used the term "digitalization" and discussed the social implications of the "digitalization of society" when considering the objections and potentials for computer-assisted humanities research. It has come to be seen as structuring different areas of social life around digital communication and media infrastructure.

In the following years, other scholars addressed this issue in their work. One of them [80] explored the technological benefits of digitalization and emphasized the importance of implementing information infrastructure in enterprises to support strategic decision-making by managers. The increasing automation of work tasks, the emergence of the Internet as a global communications infrastructure, and the development of satellites and televisions in the 1980s and 1990s were among the most important developments, and the 1990s caused companies to digitalize their businesses [4]. It was confirmed that digitalization and increased productivity are the most critical factors in explaining downsizing in companies [79].

As we enter the twenty-first century, researchers have paid more attention to the benefits and challenges that digitalization brings. Its impact has been particularly evident in the area of business model innovation, in the internationalization of business [75], and in the building of business relationships and networks [8]. At the same time, the transformation of business processes and strategy [9] and the optimization of human-technology relationships have become a focus of interest for the professional community. Scholars have focused on understanding the differential effects of digitalization on employment in different economic, production, and social contexts [16]. The impetus for this was the emergence of the first online communication programs (e.g., Skype in 2003), social networks (e.g., Facebook in 2004), and online training courses. These circumstances and the subsequent intensive development of smartphones and mobile applications have become essential milestones for digital entrepreneurship since 2013 [29].

The Fourth Industrial Revolution, or Industry 4.0, was first mentioned in 2011 in Germany as a proposal to develop an economic policy based on modern high-tech strategies to increase the manufacturing sector's competitiveness. Industry 4.0 has opened up new opportunities for organizational change. Moreover, emerging DTs started to spread much faster and on a larger scale [1]. It has manifested in manufacturing DTs using cyber-physical manufacturing processes, digital supply chains, autonomous machines, and data transfer. Overall, the importance of employees has increased as the integration of the digital and physical worlds has begun to occur [62, 73].

In 2021, the European Commission (EC) announced the start of the Fifth Industrial Revolution or Industry 5.0. The reason for this was that Industry 4.0 focused more on DTs to increase the efficiency and flexibility of production and placed less emphasis on social justice. In contrast, Industry 5.0 focuses on sustainability, resilience, and a human-centered approach by building on human needs and interests at the center of the production process [12]. The main idea is that DTs should serve people and society and be used in production with the needs of employees in mind [37]. It is also worth noting that Industry 5.0 is not a chronological continuation of Industry 4.0 or its alternative. Instead, it determines how industry and new social trends and needs will coexist, complementing and extending the characteristics of Industry 4.0 [77].

Today, digitalization is closely related to labor market conditions and DTs that affect employment outcomes [54]. The simultaneous progression of the Fourth and Fifth Industrial Revolutions increased the demands on the cognitive and creative skills of HRs and required training [42]. Digitalization brings new approaches to work and changes the employees' expectations across industries, making it managers' priority worldwide [34]. As Industry 4.0 and Industry 5.0 continue, these trends will further deepen and transform HRM in enterprises.

2.2 Definition of Terms Related to Digitalization

Digitalization is the subject of much debate among organizations, scholars, and practitioners in many countries. There are many perspectives on how to define the term. Section 2.2 will present definitions of digitization, digitalization, and digital transformation and will discuss how they have been perceived from 2009 to 2022. The phenomenon has begun to attract increasing interest from the academic community and has also been reflected in the growing number of scholarly publications on the topic (Fig. 2.1).

The search results show that there has been a doubling of publications related to digitalization since 2015. From this period to the end of 2022, the most significant contributions to research on digitalization have been made by scientists and researchers from Germany (i.e., 2764 and 4408 WoS and Scopus entries, respectively), Russia (i.e., 2331 and 4606 WoS and Scopus entries respectively), China (i.e., 1099 WoS and 1653 in Scopus, respectively), Italy (i.e., 933 WoS and 1,307 in Scopus, respectively) and the United Kingdom (UK) (i.e., 871 WoS and 1184 in Scopus, respectively).

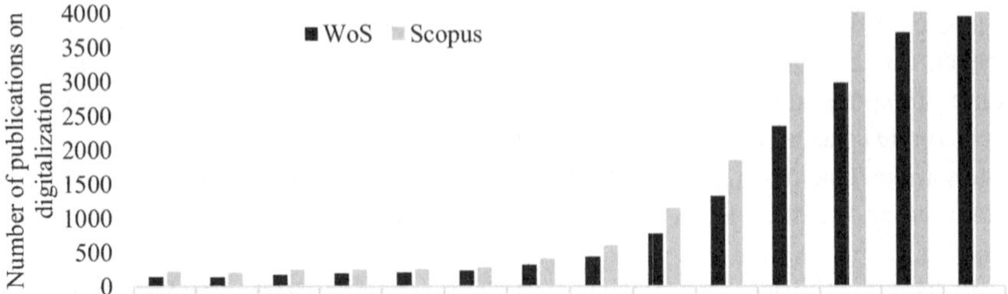

	2009	2010	2011	2012	2013	2014	2015	2016	2017	2018	2019	2020	2021	2022
WoS	143	136	172	196	204	232	318	435	771	1,31	2,34	2,97	3,71	3,94
Scopus	216	201	237	246	248	276	411	596	1,13	1,83	3,25	4,69	6,31	7,10

Fig. 2.1 Number of publications on digitalization in WoS and Scopus databases

The definitions of digitalization proposed by key Czech and foreign authors and experts are presented below.

Baraldi and Nadin [8] state that digitalization is a continuous process that combines technology-supported routine work with human interactions and adaptive communication.

Zhongjun et al. [82] approach digitalization as one component of a conceptual model for high-speed and automated e-commerce.

According to Tilson et al. [67], digitalization is a socio-technical process of applying techniques to the broader social and institutional contexts comprising DTs' infrastructure.

In his article, [48] defines digitalization as an emerging communication medium, technological convergence, and the interconnection of technologies whose functional purpose is to disseminate digitized communication.

Kagermann [30] describes digitalization as the networking of people and things and the convergence of real and virtual worlds, which is made possible by information and communication technology (ICT).

Brennen et al. [11] characterize digitalization as the adoption or increased use of computer technology or DT by organizations, industries, countries, etc.

Mancini et al. [39] view digitalization as a technological innovation that implies a different philosophy in how companies manage their business and processes.

Balcerzak and Pietrzak [5] hold the view that digitalization can become an advantage for businesses if tangible investments in the infrastructure of the digital economy, especially investments in ICT, are coupled with intangible investments in human capital and resources of companies such as new business processes or strategies and decentralized organizational structures.

Legner et al. [34] explain digitalization as the sum of diverse sociotechnical phenomena and the process of adopting and using DTs in broader individual, organizational, and social contexts.

The authors [43] agree that digitalization represents a set of values leading to the optimization of economic growth and includes an application that supports the production process.

OECD [49] defines digitalization as linking data and DTs, which leads to new activities or changes to existing ones.

Fabbri and Scapolan [21] describe digitalization as a process of organizational transformation characterized by the replacement of analogue rules (i.e., implicit and informal) for decision-making and action by digital rules (i.e., explicit and formalized).

According to Stockhinger and Teubner [63], digitalization is a complex phenomenon that encompasses social, mobile, analytics, and cloud technologies that empower organizations and the process of triggering innovation. It is possible on the condition that digitalization is linked to all organization processes, which will give access to high-quality information from daily operations to increase the value of decision-making [27].

Westerlund [75] states that digitalization implies using DTs to change a company's business model. In addition, [28] add that the expected outcomes of digitalization, in the long run, are reduced demand for HRs, increased a typical forms of employment, and a widening pay gap within occupations.

CAS [13] emphasizes that the essence of digitalization lies in a transformation process supported by technological processes using the potential of HR to the maximum extent possible.

The above definitions show that over time, the focus of researchers and practitioners has gradually shifted from the impact of digitalization on industrial growth and business productivity to questions of its impact on employment and work automation. As recently as the first decade of the twenty-first century, digitalization was approached as a transformation of business processes based on technology. It needs to be more to explain its nature, and its complex impact is no longer sufficient. According to the book's Author, digitalization should express the interdependence and inseparability of technological progress and the development potential of HRs, on which the application of DTs in business practice is highly dependent.

It should be stressed that many scientific publications and studies also use the terms "digitization" and "digital transformation" to refer to the term "digitalization". Some authors see them as synonyms and confuse their use in practice without any significant distinction. This approach is followed, for example, by [25], who aim to create a unified definition of digital transformation that should include other related concepts (digitization, digitalization, etc.). The final unified definition of digital transformation proposed by [25, p. 12] is as follows:

> A process of fundamental change, enabled by the innovative use of DTs accompanied by the strategic use of key resources and capabilities, to radically improve an organization, business network, industry or society and redefine its value proposition to its stakeholders.

However, many more authors argue that digitization, digitalization, and digital transformation are not equivalent and are distinct [34, 49, 57, 67, 71]. Rachinger et al. [53] argue that digitization, the process of converting analogue data into digital datasets, serves as a framework for digitalization, behind which lie as yet untapped digital opportunities. According to [10], digital transformation encompasses the entire organization beyond digitization and digitalization. In contrast, digital transformation is a much more complex issue, characterized by digital initiatives in terms of organizational goals, policies and culture, risk management, and employee skills, thus placing great demands on HRs, technical, and financial resources in organizations [78]. It involves converting business processes, operations, and structures to gain new technological advantages [22].

Verhoef et al. [71] view digitization, digitalization, and digital transformation as three distinct phases of the digital transformation process, one following the other. While digitization does not change the value-creating activities for the enterprise, digitalization contributes to improving production processes and customer experience, and digital transformation is inherently linked to strategic changes in the business.

According to Sen [57], digitization focuses on integrating of DTs in the workplace. At the same time, digitalization requires a complete rethinking of the organization to its corporate culture, day-to-day operations, and future development strategies. In contrast, digital transformation requires a new work environment, better employees' learning and training, and empathy for their need to discover their full potential in the digital age. It aligns with [50] notion of a multi-level digital transformation process that changes the way of working and roles at levels, including processes, organization, business domain, and society.

Maheshwari [38] holds a similar view and argues that digital transformation is a higher form of organizational adaptation to changing customer demands by applying DT to improve all business segments significantly. Moreover, [35] emphasizes that the process of transferring the core business operations of companies to an online environment or otherwise computer-driven medium in digital transformation is long-term and cannot be fully completed due to the continued expansion of new DTs.

The presented review of the current literature shows that, according to several authors, digitization, digitalization, and digital transformation are distinct concepts, each of which focuses to some extent on technical aspects yet has its specific meaning. In contrast, experts believe that all these terms represent different sides of the same process and should be included in a standard definition. The consequence is the need for a unified approach to the definition of digitization in Czech and foreign scientific literature, which points to a gap in the existing knowledge. Because of this, the Author of the book proposes linking the concepts of digitization, digitalization, and digital transformation into a single idea of digitalization. To this end, the Author of the book presents her comprehensive definition below.

Definition. "Digitalization with HRM is the integration of new DTs into existing company processes to increase efficiency. Its level is characterized by the readiness of HRs to regularly develop and educate themselves on the job in response to current digital trends in the business environment".

In the following chapters, the book's Author will use the term digitalization in the above-mentioned meaning to elaborate further on this book.

2.3 Digital Technologies as Tools for Digitalization

The twenty-first century is characterized by ongoing digitalization, a complex and multifaceted phenomenon. According to the Council of the EU, more jobs are increasingly subject to digitalization, and technology plays a significant role in developing more flexible work environments that adapt to the needs of highly mobile societies [17]. Originally used to perform routine tasks in the workplace, DTs currently support maintaining social connections and building community in organizations [7].

It should be emphasized that the power of DTs does not lie in their technical characteristics but stems from how companies integrate them to transform their business [31]. Moreover, the successful integration of digital business processes depends on the ability of employees to quickly learn how to use new DTs in a work context and to share lessons learned with colleagues [6]. Consequently, employees must acquire relevant skills such as adaptability, critical thinking and decision-making, advanced digital literacy, complex problem-solving, and advanced data analysis [42].

The latest DTs are crucial for implementating and applying digitalization and the modernization of socio-technological relations between man and machine within and between organizations [21]. According to [9], DTs can be defined as a combination of information technology (IT) and computing, communication, and networking technologies that fundamentally change businesses' processes and strategies, products and services, and inter-organizational relationships in business networks.

As [31] argue, DTs are also considered social, mobile, analytics and cloud technologies. Vial [72] adds the Internet of Things (IoT) and uses an acronym that combines both terms—SMACIT. Ministry of Education [44] uses the term DTs to refer to a combination of technical tools and devices of a material nature and technical processes that are not material and are intended for the collection, transmission, storage, and processing of data through information and communication channels.

Some other authors [20, 36, 41, 46, 59, 64] provide definitions of the different components of DTs, which are summarized in Table 2.1.

According to [32], DTs can also be understood as emerging technologies, which are cyber-physical systems, IoT, cloud computing, Big data, Artificial Intelligence (AI), Virtual Reality (VR), and Augmented reality. In this book, the Author leans towards the definition of [32] as it accurately captures the subject matter and identifies specific elements of DTs currently used in enterprises.

It should be emphasized that the authors of scientific works disagree on differing typologies of DTs. Spremic [61] states that all DTs can be divided into two basic types in practice:

- Primary DTs: already mature established, and commonly used in almost every business like mobile, social, cloud, big data, sensors, and IoT.
- Secondary DTs: can still be developed and improved but have the potential to be used in business like additive manufacturing (i.e., 3D printers), robotics, drones, wearable technologies, holograms, virtual and augmented reality, cognitive technologies (e.g., AI), deep learning algorithms, facial and speech recognition.
- Pousttchi et al. [51] divide DTs differently into the following categories:

Table 2.1 Theoretical definition of different types of DTs according to selected authors

Types of DTs	Author	Definition
AI	[20]	Machine learning based on the simulation of human behavior using computer systems, applications, and algorithms capable of independent decision-making and problem-solving based on collected data
IoT	[64]	Connecting to intelligent assets via an open network that is complemented by advanced technologies such as scalable computing, information management, analytics, and mobility
Cloud technologies	[41]	Model enabling convenient on-demand network access to a shared pool of computing resources (e.g., servers, storage, applications, and services) that can be rapidly provisioned with minimal management effort or service provider interaction
Big data	[46]	A set of available and relevant data from different sources, characterized by its diversity, large volume, and speed of creation, aggregation, and transmission
VR	[36]	A set of human–computer interaction technologies that enable the creation of computer-generated virtual environments in which users can freely interact with virtual objects, agents, or other users
Augmented reality	[59]	A technology that transforms physical environments through overlapping virtual elements (i.e., text, images, videos, or digital projects added by users) that are distinctive in their novelty, interactivity, and liveliness

- Communication and other supporting technologies: provide the basis for developing of complex systems (i.e., mobile communication systems, auto-identification systems, additive manufacturing, and computer architecture).
- Technologies combining hardware and software in intelligent systems: mobile devices, stationary devices, ubiquitous computing, human–computer interface, technical enhancement of the human body and mind, human enhancement, robotics, and IoT functionality.
- Data technologies: established database technologies, new database technologies, IT infrastructure, data analytics, big data, AI, and information security.

The assessment of the impact of DTs on business operations has been the subject of research by many authors of scientific publications. Tsou and Chen [68] confirm that using DTs positively affects digitalization strategy and organizational innovation, which affect business performance.

Moreover, [14] argue that DT only positively impacts the internationalization of e-commerce if it is embedded in business processes, organizational innovation, and investments have been made in employees' digital competencies.

Shpak et al. [60] point out that by digitizing routine tasks and making them DT and intelligent systems, HRs in enterprises can be spent on complex and more strategically significant activities. Pousttchi et al. [51], in their research on the impact of DTs on companies' business processes, identify 60 possible effects that can be

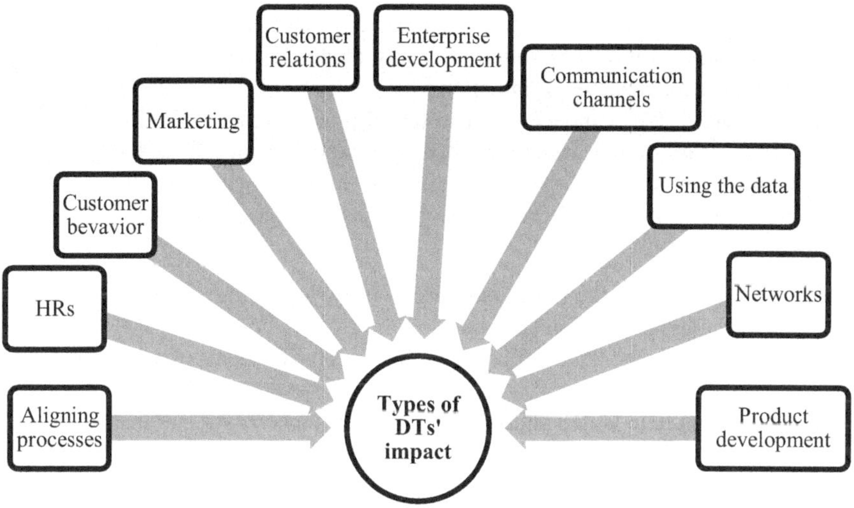

Fig. 2.2 Types of impact of digitalization through DTs [51]

distinguished along three dimensions (i.e., value creation, value proposition, and customer interaction) and can be categorized into basic types shown in Fig. 2.2.

Regarding enterprise development, digitalization can increase or mitigate dependence on market participants and structural circumstances, changing business models, organizational structures, and outsourcing strategies. In addition, digitalization introduces new recruitment practices, helps address the shortage of skilled employees and promotes the participation of HRs in the company's decision-making processes.

The Unified Theory of Acceptance and Use of Technology (UTAUT) model can be used to understand the impact of DT from an organizational perspective. This model was initially developed by [70] to explain and predict technology acceptance in an organizational context. The authors assumed that the perceived likelihood of DT adoption depends on the four constructs' direct influence. These are:

- Expected performance: the extent to which the individual believes that using the system will help them achieve higher job performance.
- Expected effort: the level of ease associated with using the system.
- Social influence: the degree to which an individual perceives that it is vital that others believe they can us e the new system.
- Facilitating conditions: the extent to which the individual believes that there is an organizational and technical infrastructure to support the use of the system.

It should be noted that the influence of predictors is moderated by age, gender, experience, and voluntariness of use, which determine their strength and intent. Age moderates the effect of all four predictors. Gender moderates the relationships between expected effort, expected performance, and social influence. Experience moderates the strength of the relationships between expected effort, social influence, and facilitating conditions. Voluntary use only moderates the relationship between social influence and behavioral intention [70].

The original UTAUT model was also tested in non-organizational settings over the next ten years, found widespread use, and expanded into the UTAUT2 model in 2012 to be applied to new technologies and user segments [70, 76]. UTAUT2 assumes that the influence of three additional constructs supports individuals' use of technology. They are:

- Hedonic motivation: the fun or pleasure of using technology, which is important in determining its adoption and use.
- Value for money: the trade-off between the perceived benefits of applications and their financial cost.
- Habit: the extent to which people tend to perform a behavior automatically.

The UTAUT and UTAUT2 models have been tested in different countries with different cultures, economies, and levels of technology diffusion, as well as in various industries, such as mobile banking [47].

As DTs are developing rapidly, their use leads to qualitative radical changes in the functioning of businesses, requiring HRs to be ready for digitalization and overall increasing the competitiveness of both companies and individual employees. The importance of DTs, which simultaneously connect all chains of product creation, production operation, and services, will continue to grow, increasing the demands on employee competencies necessary to use these DTs at work.

2.4 Impact of the COVID-19 Pandemic on Digitalization in Enterprises

As the COVID-19 pandemic's impact on digitalization in enterprises has yet to be sufficiently explored [2], Sect. 2.4 summarises the main findings of the research already conducted in this area. In particular, the coronavirus epidemic has become an unprecedented catalyst for change, capturing businesses worldwide in almost every industry sector and prompting public and private institutions to invest in digitalization at a pace never before experienced [56]. It is confirmed by a survey conducted in June 2020 among 2,569 senior executives of large enterprises in Australia, France, Germany, Italy, Japan, Singapore, Spain, the UK, and the US [69]. Industries that have experienced a significantly accelerated pace of digitalization due to the ongoing coronavirus pandemic are presented in Fig. 2.3.

Regarding the Czech companies' approach to digitalization, the tendency to accelerate the digitalization of activities rather than wait to see how the situation around the COVID-19 pandemic develops prevails in the construction (11.0 versus 9.0%) and finance and insurance (24.0 versus 7.0%) sectors. In contrast, retail businesses (4.0 versus 11.0%) and manufacturing and automotive industries (8.0 versus 13.0%)

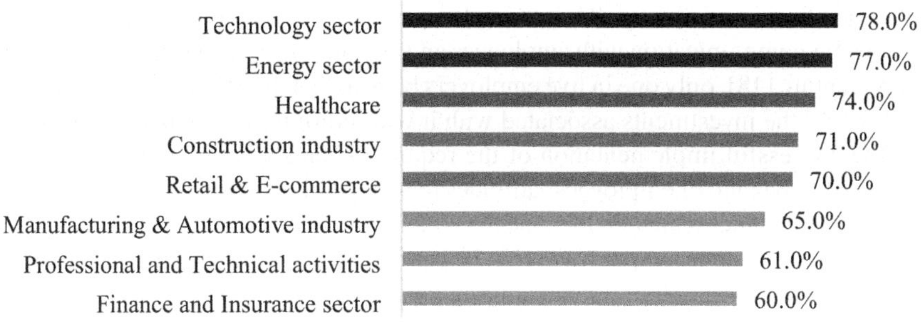

Fig. 2.3 Industries with accelerated digitalization as a result of the COVID-19 pandemic [69]

prefer to wait until they decide to digitalize further in these ambiguous conditions. It is highlighted by the results of a survey [40] summarizing data from 26,130 employers in six industries in 43 countries. It is important to note that, in an international comparison, the CR, like several other EU countries (i.e., Portugal, Spain, Slovenia, and France), is still digitalizing the least [40]. The leader is Germany, where companies plan to digitalize even more than in the US (44.0 versus 13.0%).

The speed of DT adoption has left organizations needing more time to adapt their structures, processes, and culture to the new work environment [55]. Reluctance to change organizational routines, lack of awareness of the latest technologies, the potential benefits of digitalization, and the security measures associated with the digitalization of business processes have become severe barriers to adopting DTs in the pandemic period. The opportunities offered by DTs and the COVID-19 pandemic were limited by the inflexibility of enterprises [2]. Barriers have also arisen due to the need for a clear digitalization strategy and the lack of staff competencies to support its smooth progress [69]. The most recent survey of [52] found that 20,000 employees in 44 countries are concerned that their employers are not receiving sufficient training in digital and technology skills. Teltscher [66, p. 1] states:

> The pandemic has spurred innovation in digital tools and platforms and increased the digitalization of business processes, products, and services. As the pandemic continues, there is an increasing emphasis on the use of digital technologies.

Following this, [19] highlights the role of HR professionals in advancing corporate objectives during the pandemic:

> COVID has forced every organization to undertake a deep self-reflection. HR leaders, who were initially driven by the need to protect their employees from the virus, are now addressing longer-term fundamental questions and practices around how best to manage and support employees, how best to collaborate, with what technologies, and what basic ways of working are best for both productivity and employee wellbeing.

The subsequent need to adapt to the new working conditions caused by the COVID-19 pandemic leads to a rethinking of HR T&D strategies [81] and differentiated requirements for HR managers in different sectors of the economy [33]. Suddenly, companies whose business could be run digitally have moved to a remote or home office environment. They started new virtual collaboration tools and applications for communication with employees and customers [3]. According to research by Accenture [18], only one in five employers believe that the HR T&D strategy used to date and the investments associated with it will return to pre-pandemic levels.

The successful implementation of the required changes is highly dependent on both the knowledge of employees' attitudes towards digitalization and the expansion of their competencies, and on the managers' ability to approach and promote change as a complex adaptation process [13]. In particular, 2.4 million unfilled positions were expected to remain in manufacturing companies between 2018 and 2028 due to a lack of employee skills that support introducing and using new technologies. These skills include critical thinking, programming, and digital skills [24].

In conclusion, HR professionals can inspire others in the organization to use digital ways of working and communicating, which have already proven their worth in the context of the coronavirus pandemic. However, some challenges lie ahead for organizations engaged in digitalization in a pandemic. These include the need to conduct data analytics, implement digital recruitment, provide staff T&D development, and retain talent as hybrid forms of working and remote working increase. These challenges should be carefully addressed.

The findings from this chapter indicate that digitalization is an ongoing and evolving process in which more and more businesses of different sectors and sizes are becoming involved. Digitalization is a general organizational phenomenon relevant to almost all organizational areas. An unforeseen pandemic spurred the recent intensification of the use of DTs in business processes. Even during the pandemic, some scholars and researchers pointed out that with further technological advances, humans would not be fully represented by software or technology in the workplace. On the contrary, the role and importance of HRs will continue to increase, and the further successful digitalization of business processes will depend on them.

References

1. Albano, R., Curzi, Y., Fabbri, T.: Work autonomy, control and discretion in Industry 4.0. In: Cantoni, F., Mangia, G. (eds.) Human resource management and digitalization, pp. 95–113. Routledge, Abingdon, New York, Torino (2019)
2. Amankwah-Amoah, J., Khan, Y., Wood, G.: COVID-19 and digitalization. J. of Bus. Res **136**, 602–611 (2021). https://doi.org/10.1016/j.jbusres.2021.08.011
3. ASME CR, IPSOS: Inovace v malých a středních firmách. Závěrečná zpráva 9/2020. (Innovation in small and medium-sized enterprises. Final Report 9/2020). Association of Small and Medium-sized Enterprises and Crafts of the Czech Republic. https://amsp.cz/wp-content/uploads/2020/10/IPSOS-pro-AMSP_Inovace-v-MSP_09_2020-WEB.pdf (2020). Accessed 12 June 2024
4. Ayhan, B.: Digital literacy. In: Ayhan, B. (ed.) Digitalization and Society, pp. 29–47. Peter Lang, Frankfurt am Main (2017)
5. Balcerzak, P., Pietrzak, B.: Digital economy in Visegrad countries: Multiple criteria decision analysis at regional level in the years 2012 and 2015. J. Com. **9**, 5–18 (2017). https://doi.org/10.7441/joc.2017.02.01

6. Balsmeier, B., Woerter, M.: Is this time different? How digitalization influences job creation and destruction. Res. Pol. **48**, 1–10 (2019). https://doi.org/10.1016/j.respol.2019.03.010
7. Baptista, J., Stein, M., Klein, S.: Digital work and organisational transformation: Emergent digital/human work configurations in modern organisations. J. Strat. Inf. Syst. **29**, 101618 (2020). https://doi.org/10.1016/j.jsis.2020.101618
8. Baraldi, E., Nadin, G.: The challenges in digitizing business relationships. The construction of an IT infrastructure for a textile-related business network. Techn. **26**, 1111–1126 (2006). https://doi.org/10.1016/j.technovation.2005.09.016
9. Bharadwaj, A., El Sawy, O., Pavlou, P.: Digital business strategy: Towards a next generation of insights. Mis Quart. **37**, 471–482 (2013). https://doi.org/10.25300/MISQ/2013/37:2.3
10. Bikse, V., Lusena-Ezera, I., Rivza, P.: The development of the digital transformation and relevant competencies for employees in the context of the impact of the COVID-19 pandemic in Latvia. Sust. **13**, 9233 (2021). https://doi.org/10.3390/su13169233
11. Brennen, J., Kreiss, D.: Digitization. In: Craig, T. (ed.) The International Encyclopedia of Communication Theory and Philosophy, pp. 556–566. Wiley Blackwell, Chichester (2016)
12. Breque, M., De Nul, L., Petridis, A.: Industry 5.0: Towards a sustainable, human-centric and resilient European industry. Publications Office of the EU. https://doi.org/10.2777/308407 (2021). Accessed 21 June 2024
13. CAS: Metodika pro řízení změn při digitální transformaci—digital change management (Methodology for change management in digital transformation—digital change management). Czech Agency for Standardization. https://www.koncepcebim.cz/wp-content/uploads/2023/04/Metodika-pro-rizeni-zmen-pri-digitalni-transformaci-Digital-Change-Management_Agentura-CAS.pdf (2020). Accessed 19 June 2024
14. Casseta, E., Monarca, U., Dileo, I., Di Berardino, C., Pini, M.: The relationship between digital technologies and internationalisation: Evidence from Italian SMEs. Ind. Inn. **27**, 311–339 (2020). https://doi.org/10.1080/13662716.2019.1696182
15. Ceruzzi, P.E.: A History of Modern Computing. MIT Press, Cambridge, London (2003)
16. Cirillo, V., Evangelista, R., Guarascio, D., Sostero, M.: Digitalization, routineness and employment: An exploration on Italian task-based data. Res. Pol. **50**, 1–35 (2019). https://doi.org/10.1016/j.respol.2020.104079
17. Council of the EU: Council recommendation of 22 May 2018 on key competences for lifelong learning (text with EEA relevance). Off. J. Eur. Un. **61**, C189/01 (2018)
18. Crowley, E., Overton, L.: Learning and skills at work survey 2021. Chartered Institute of Personnel and Development. https://www.cipd.co.uk/Images/learning-skills-work-report-2021-1_tcm18-95433.pdf (2021). Accessed 17 June 2024
19. Crux, R., Brantingham, L., Elmore, P.: What impact has COVID had on HR practices and what can we learn from?. NGS Global. https://www.ngs-global.com/images/PDF/RCPELB-COVID-HR-Practices.pdf (2021). Accessed 18 June 2024
20. Donkor, C., Slobodjanjuk, A., Kremer, K., Weisshaar, J.: The way we work—in 2025 and beyond. PwC. https://www.pwc.ch/en/publications/2017/the-way-we-work-hr-today_pwc-en_2017.pdf (2017). Accessed 22 May 2024
21. Fabbri, T., Scapolan, A.C.: Digitalization and HR analytics: A big game for an HR manager. In: Cantoni, F., Mangia, G. (eds.) Human Resource Management and Digitalization, pp. 243–254. Routledge, Abingdon, New York, Torino (2019)
22. Fenech, R., Baguant, P., Ivanov, D.: The changing role of human resource management in an era of digital transformation. J. Man. Inf. Dec. Sci. **22**, 166–175 (2019)
23. Fierro, P., Briganti, P., Varriale, L.: Individual versus organizational learning for knowledge in the innovation 4.0 era. In: Cantoni, F., Mangia, G. (eds.) Human resource management and digitalization, pp. 153–176. Routledge, Abingdon, New York, Torino (2019)
24. Giffi, C., Wellener, P., Dollar, B., Manolian, H.A., Monck, L., Moutray, C.: Deloitte skills gap and future of work in manufacturing study 2018. Deloitte Development LLC. https://www2.deloitte.com/content/dam/insights/us/articles/4736_2018-Deloitte-skills-gap-FoW-manufacturing/DI_2018-Deloitte-skills-gap-FoW-manufacturing-study.pdf (2018). Accessed 17 Apr 2024

25. Gong, C., Ribiere, V.: Developing a unified definition of digital transformation. Techn. **102**, 102217 (2021). https://doi.org/10.1016/j.technovation.2020.102217
26. Hartree, D.: The ENIAC, an electronic calculating machine. Nat. **157**, 527 (1946). https://doi.org/10.1038/157527a0
27. Hellsten, P., Paunu, A.: Digitalization: A concept easier to talk than to understand. In: Salgado, A., Bernardino, J., Filipe, J. (eds.) Proceedings of the 12th International Joint Conference on Knowledge Discovery, Knowledge Engineering and Knowledge Management (IC3K 2020), pp. 226–233. Science and Technology Publications, Setúbal (2020)
28. Ivanová, E., Žárská, V., Masárová, J.: Digitalization and human capital development. Entr. Sust. Iss. **9**, 402–415 (2021). https://doi.org/10.9770/jesi.2021.9.2(26)
29. Jindra, J., Kubátová, P., Klíma, K., Špáčková, K., Šťáfková, M., Šťáfek, P.: Analýza existujících profilů digitálních kompetencí a standardů poskytovatelů vzdělávání v oblasti digitálních kompetencí (Analysis of existing digital competence profiles and standards of digital competence education providers). Ministry of Labour and Social Affairs of the Czech Republic. https://media.portaldigi.cz/a835d570-5a39-4cac-829e-e90c3250c9a5 (2020). Accessed 21 June 2024. Accessed 18 May 2024
30. Kagermann, H.: Change through digitization — value creation in the age of Industry 4.0. In: Albach, H., Meffert, H., Pinkwart, A., Reichwald, R. (eds.) Management of Permanent Change, pp. 23–45. Springer Gabler, Wiesbaden (2015)
31. Kane, G., Palmer, D., Phillips, A.N., Kiron, D., Buckley, N.: Strategy, not technology, drives digital transformation. Becoming a digitally mature enterprise. MIT Sloan Management Review, Deloitte University Press. https://www2.deloitte.com/content/dam/Deloitte/fr/Documents/strategy/dup_strategy-not-technology-drives-digital-transformation.pdf (2015). Accessed 5 Apr 2024
32. Khin, S., Ho, T.: Digital technology, digital capability and organizational performance: A mediating role of digital innovation. Int. J. Inn. Sci. **11**, 177–195 (2019). https://doi.org/10.1108/IJIS-08-2018-0083
33. Kolot, A., Lopushniak, H., Kravchuk, O., Varis, I., Ryabokon, I.: Transferable competencies of HR managers under global socio-economic changes. Prob. Persp. Man. **20**, 322–341 (2022). https://doi.org/10.21511/ppm.20(1).2022.27
34. Legner, S., Eymann, T., Hess, T.: Digitalization: Opportunity and challenge for the business and information systems engineering community. Bus. Inf. Syst. Eng. **59**, 301–308 (2017). https://doi.org/10.1007/s12599-017-0484-2
35. Levit, A.: Humanity Works: Merging Technologies and People for the Workforce of the Future. Kogan Page, London, New York (2019)
36. Li, C., Shing IP, H.: Defining virtual reality enabled learning. Int. J. Inn. Learn. **31**, 291–306 (2022). https://doi.org/10.1504/IJIL.2022.122128
37. Lu, Y., Sastre Adrados, J., Shivneel Chad, S., Wang, L.: Humans are not machines—anthropocentric human-machine symbiosis for ultra-flexible smart manufacturing. Engineering **7**, 734–737 (2021). https://doi.org/10.1016/j.eng.2020.09.018
38. Maheshwari, A.: Digital Transformation: Building Intelligent Enterprises. Willey, Hoboken (2019)
39. Mancini, D., Lamboglia, R., Castellano, N.G., Corsi, K.: Trends of digital innovation applied to accounting information and management control systems. In: Corsi, K., Castellano, N.G., Lamboglia, R., Mancini, D. (eds.) Reshaping Accounting and Management Control Systems, pp. 1–19. Springer, Cham (2017)
40. ManpowerGroup: Revoluce dovedností: Restart. Dopad COVID-19 na digitalizaci a dovednosti: Nová budoucnost práce (Skills Revolution: reboot. The impact of COVID-19 on digitalisation and skills: the new future of work). https://www.manpowergroup.cz/wp-content/uploads/2021/03/ManpowerGroup_Revoluce_dvoednost_2021.pdf (2020). Accessed 12 Apr 2024
41. Mell, P., Grance, T.: The NIST definition of cloud computing. Recommendations of the National Institute of Standards and Technology. Special publication 800-145. U.S. Department of Commerce. https://nvlpubs.nist.gov/nistpubs/Legacy/SP/nistspecialpublication800-145.pdf (2011). Accessed 28 Apr 2024

42. Melnyk, L., Kubatko, O., Matsenko, O., Balatskyi, Y., Serdyukov, K.: Transformation of the human capital reproduction in line with Industries 4.0 and 5.0. Prob. Persp. Man. **19**, 480–494 (2021). https://doi.org/10.21511/ppm.19(2).2021.38
43. Milošević, N., Dobrota, M., Rakočević, R.: Digital economy in Europe: Evaluation of countries' performances. Proc. Fac Econ. Rej. **36**, 861–880 (2018). https://doi.org/10.18045/zbefri.2018.2.861
44. Ministry of Education, Youth and Sports of the Czech Republic: Digitální kompetence. Pojetí tematické oblasti v projektu P-KAP. IV. část verze pro decizní sféru vzdělávácí politiky České republiky (Digital competence.The concept of thematic area in the P-KAP project.Part IV of the version for the education policy sphere of the Czech Republic). https://archiv-nuv.npi.cz/uploads/P_KAP/ke_stazeni/pojeti_decizni_sfera/AKTUALIZACE_2020/2020_Digitalni_kompetence_IV_podrobne_pojeti_oblasti_intervence_aktualizace.pdf (2020). Accessed 12 May 2024
45. Moore, G.: Cramming more components onto integrated circuits. El. **38**, 1–4 (1965)
46. Moorthy, J., Lahiri, R., Biswas, N., Sanyal, D., Ranjan, J., Nananth, K., Ghosh, P.: Big data: Prospects and challenges. Vikalpa **40**, 74–96 (2015). https://doi.org/10.1177/0256090915575450
47. Mütterlein, J., Kunz, R., Baier, D.: Effects of lead-usership on the acceptance of media innovations: A mobile augmented reality case. Techn. For. Soc. Chan. **145**, 113–124 (2019). https://doi.org/10.1016/j.techfore.2019.04.019
48. Ocampo, S.: Digitalisation and technological convergence from the sociological perspective of general social systems theory. Rev. Gest. De las Pers. y Tech. **5**, 153–165 (2012)
49. OECD: Going digital in a multilateral world — an interim report to ministers. Organization for Economic Development and Co-operation, Directorate for Science, Technology and Innovation, Committee on Digital Economy Policy. https://one.oecd.org/document/C/MIN(2018)6/en/pdf (2018). Accessed 5 May 2024
50. Parviainen, P., Tihinen, M., Käriäinen, J.: Tackling the digitalization challenge: How to benefit from digitalization in practice. Int. J. Inf. Syst. Proj. Manage. **5**, 63–77 (2017). https://doi.org/10.12821/ijispm050104
51. Pousttchi, K., Gleiss, A., Buzzi, B., Kohlhagen, M.: Technology impact types for digital transformation. In: Becker, J., Novikov, D. (eds.) Proceedings of the 21st Conference on Business Informatics CBI 2019, vol. 2, pp. 487–494. Institute of Electrical and Electronics Engineers, Piscataway (2019)
52. PWC: Global workforce hopes and fears survey 2022. https://www.pwc.com/gx/en/hopes-and-fears/downloads/global-workforce-hopes-and-fears-survey-2022-v2.pdf (2022). Accessed 18 May 2024
53. Rachinger, M., Rauter, R., Müller, C.: Digitalization and its influence on business model innovation. J. Man. Techn. Man. **30**, 1143–1160 (2019). https://doi.org/10.1108/JMTM-01-2018-0020
54. Reljic, J., Evangelista, R., Pianta, M.: Digital technologies, employment and skills. Ind. Corp. Chang. dtab059:1–24 (2021). https://doi.org/10.1093/icc/dtab059
55. Reuschl, A.J., Deist, M.K., Maalaoui, A.: Digital transformation during a pandemic: Stretching the organizational elasticity. J. Bus. Res. **144**, 1320–1332 (2022). https://doi.org/10.1016/j.jbusres.2022.01.088
56. Savastano, M., Zentner, H., Spremić, M., Cucari, N.: Assessing the relationship between digital transformation and sustainable business excellence in a turbulent scenario. Tot. Qual. Man. 1–23 (2022). https://doi.org/10.1080/14783363.2022.2063717
57. Sen, S.: Digital HR strategy: Achieving Sustainable Transformation in the Digital Age. Kogan Page, New York (2020)
58. Shannon, C.: A mathematical theory of communication. Bell Syst. Techn. J. **27**, 379–423 (1948). https://doi.org/10.1002/j.1538-7305.1948.tb01338.x
59. Shi, M., Zhang, C., Hu, J.: An innovative marketing strategy: Adopting AR VR in jewelry sales. In: Patnaik, S. (ed.) Proceedings of the 2nd International Conference on Applied Machine Learning ICAML 2020, pp. 333–338. IEEE Computer Society, Los Alamitos (2020)

60. Shpak, N., Maznyk, L., Dvulit, Z., Seliuchenko, N., Dragan, O., Doroshkevych, K.: Influence of digital technologies on the labor market of HR specialists. In: Vysotska, Y., Kupriianov, O., Kanishcheva, O., Cherednichenko, O., Hamon, T., Gradar, N. (eds.) Proceedings of the 6th International Conference on Computational Linguistics and Intelligent Systems COLINS 2022, pp. 1475–1487. CEUR-WS, Gliwice (2022)
61. Spremic, M.: Governing digital technology—how mature IT governance can help in digital transformation? Int. J. Econ. Man. Syst. **2**, 214–223 (2017)
62. Stocker, A., Brandl, P., Michalczuk, R., Rosenberger, M.: Human-centered ICT tools for smart factories. El. Inf. **131**, 207–211 (2014). https://doi.org/10.1007/s00502-014-0215-z
63. Stockhinger, J., Teubner, A.: How digitalization drives the IT/IS strategy agenda. University of Münster, European Research Center for Information Systems. Available at https://www.econstor.eu/bitstream/10419/213281/1/1687593957.pdf (2019). Accessed 30 May 2024
64. Šimec, A., Ibrahimov, M., Aleksic, A., Dukic, D.: Cyber-attacks and internet of things as a threat to critical infrastructure. In: Šimec, A., Ibrahimov, M., Aleksic, A., Dukic, D. (eds.) Proceedings of the 37th International Scientific Conference on Economic and Social Development, pp. 108–111. Vazardin Development and Entrepreneurship Agency, Baku (2019)
65. Tabak, D.: Digitalisation of control systems. Comp. Aid. Des. **3**, 13–18 (1971). https://doi.org/10.1016/0010-4485(71)90063-7
66. Teltscher, S.: Introduction. In: Mcdonald, M. (ed.) Digital Skills Insights 2020, pp. 1–9. International Telecommunication Union, Geneva (2020)
67. Tilson, D., Lyytinen, K., Sørensen, C.: Digital infrastructures: The missing IS research agenda. Inf. Syst. Res. **21**, 748–759 (2010). https://doi.org/10.1287/isre.1100.0318
68. Tsou, H., Chen, J.: How does digital technology usage benefit firm performance? Digital transformation strategy and organisational innovation as mediators. Techn. An. Strat. Man. (2021). https://doi.org/10.1080/09537325.2021.1991575
69. Twilio: COVID-19 digital engagement report. https://pages.twilio.com/rs/294-TKB-300/images/UPDATE_Aug_Twilio_COVID-19_Digital_Engagement_Report.pdf (2020). Accessed 10 Apr 2024
70. Venkatesh, V., Morris, M., Devis, G.: User acceptance of information technology: Toward a unified view. MIS Quat. **27**, 425–478 (2003). https://doi.org/10.2307/30036540
71. Verhoef, P., Broekhuizen, T., Bart, Y.: Digital transformation: A multidisciplinary reflection and research agenda. J. Bus. Res. **122**, 879–901 (2021). https://doi.org/10.1016/j.jbusres.2019.09.022
72. Vial, G.: Understanding digital transformation: A review and a research agenda. J. Strat. Inf. Syst. **28**, 118–144 (2019). https://doi.org/10.1016/j.jsis.2019.01.003
73. Vite, C., Morbiducci, R.: Optimizing the sustainable aspects of the design process through building information modeling. Sust. **13**, 3041 (2021). https://doi.org/10.3390/su13063041
74. Wachal, R.: Humanities and computers: A personal view. North Amer. Rev. **256**, 30–33 (1971)
75. Westerlund, M.: Digitalization, internationalization and scaling of online SMEs. Techn. Inn. Man. Rev. **10**, 48–57 (2020). https://doi.org/10.22215/timreview/1346
76. Williams, M., Rana, N., Dwivedi, Y.: The unified theory of acceptance and use of technology (UTAUT): A literature review. J. Ent. Inf. Man. **28**, 443–488 (2015). https://doi.org/10.1108/JEIM-09-2014-0088
77. Xu, X., Lu, Y., Vogel-Heuser, B., Wang, L.: Industry 4.0 and Industry 5.0—inception, conception and perception. J. Man. Syst. **61**, 530–535 (2021). https://doi.org/10.1016/j.jmsy.2021.10.006
78. Yucel, S.: Estimating the benefits, drawbacks and risk of digital transformation strategy. In: Guerro, G.E. (ed.) Proceedings of IEEE International Conference on Computational Science and Computational Intelligence 2018, pp. 233–238. Institute of Electrical and Electronics Engineers, Piscataway (2018)
79. Yoo, H., Mody, B.: Predictions of downsizing in the U.S. local telephone industry. Inf. Soc. **16**, 23–33 (2000)
80. Young, J.: Benefits of digitization and the impact of computer philosophies: Reaping the rewards of information. Int. J. Com. Appl. Techn. **1**, 10–14 (1988). https://doi.org/10.1504/IJCAT.1988.062642

81. Zavyaová, E., Sokolov, D., Kucherov, D., Lisovskaya, A.: The digitalization of the human resource management: Present and future. For. STI Gov. **16**, 42–51 (2022)
82. Zhongjun, L., Lijuan, H., Qihai, Z.: The reconstruction, exploration and innovation of logos of e-commerce. In: Yu, F., Luo, Q., Chen, Y., Chen, Z. (eds.) Proceedings of the International Symposium on Electronic Commerce and Security ISECS 2008, pp. 792–795. IEEE Computer Society, Los Alamitos (2008)

Chapter 3
Theoretical Background of Digital Human Resource Management

Abstract Chapter 3 presents an overview of the historical development of the human resource management concept in the twenty-first century, emphasizing the currently ongoing Fourth and Fifth Industrial Revolutions and the emergence of digital human resource management. The individual sections of the Chap. 3 identify critical trends in the digitalization of human resource management and outline the main advantages and disadvantages of using digital technologies in particular human resource processes.

3.1 Historical Development of Human Resource Management in the Context of Industrial Revolutions

In its classical conception, HRM originated in the 1980s as an alternative to traditional personnel management forms. It was prompted by the introduction of the Harvard framework by American academics Fombrun and Beer [2].

In academic publications, HRM, human capital management, personnel management, people management, and talent management are sometimes used as synonyms [35], with which the Author of this book disagrees. This is because human capital refers to a narrower view of HRs than HRM. Moreover, as stated by [46], personnel management focuses on the employees themselves, whereas HRM deals with the overall HR needs of the organization.

The evolution of the HRM concept took time and became a natural consequence of structural changes in the world economy, production systems, and processes over decades. References to the First Industrial Revolution date back from the eighteenth to the nineteenth century. Industrialization, notably the transition to factories and mass production first in Great Britain and gradually in other Western countries, was accompanied by long working hours and unfavorable working conditions for the working classes [42].

Advances in energy technology and innovations in transport and communication characterized the Second Industrial Revolution. The world economy enabled businesses to increase production and make organizational changes based on better

A. Mazurchenko, *The Impact of the Digitalization on the Human Resource Management*, Information Systems Engineering and Management 25,
https://doi.org/10.1007/978-3-031-76790-6_3

technological equipment, allowing large companies to employ thousands of people. Between 1870 and 1914, the need for businesses to start regularly looking after their employees became clear. The demand for highly skilled employees with technical skills increased significantly [27].

The next step in the historical development of HRM was Hawthorne's study in the late 1920s and early 1930s. The original aim of this study was to demonstrate the impact of the physical work environment on employee productivity. However, it was shown that social relationships, not the physical environment, influence organizational performance [53].

Personnel management originated in the US in the 1940s and 1950s, covering the period during and after World War II. In the following decade, developmental management processes, systematic training, HR planning, and employee performance appraisal were introduced [10]. One of the main objectives of HR managers is to process and analyze information about employees and automate daily HR activities [43].

This phase saw the development of human capital theory, in which individuals' knowledge and skills began to be perceived as valuable assets of organizations [5].

The shift to electronics, IT, and production automation during the Third Industrial Revolution between 1960 and 1989 required extensive retraining of rank-and-file employees, managers, and entrepreneurs at the professional level [42].

Between 1980 and 1982, the US economy was hit by a recession, increased product market competition from Japan, and the introduction of new technologies into the manufacturing process, reducing the workforce. At the same time, in the UK, the government passed new legislation that specified working hours for women and children, set minimum wages, and protected employees from working in health-threatening environments [4].

In response, HR managers were forced to change from a routine approach to managing HR functions to a strategic one to adapt to and prepare HRs for the changes. In addition, HR services expanded to include professional counseling and became more focused on the needs of individual employees [43].

Since the early 1990s, HR activities have focused on talent management and implementing electronic systems to support recruitment, T&D, performance management, and employee compensation [48].

According to the Author of this book, all three industrial revolutions together have brought about significant technological changes, redefining the nature of work and the employee's role in most organizational processes. These revolutions have been the fundamental forces behind the evolution and development of HRM processes.

It is important to note that although the views of some scholars and practitioners on the historical development of HRM in the twentieth century are complementary, there is some inconsistency in defining time horizons for each phase of the concept's formation [1, 23, 26]. The results of the comparison are presented in Table 3.1.

The ensuing Fourth and Fifth Industrial Revolutions, which have emerged and continue steadily into the twenty-first century, are changing the nature of human work, creating the conditions for the provision of new flexible employment arrangements and the introduction of a hybrid model of work. At the same time, a socio-technical

Table 3.1 Historical development of HRM in the twentieth century, according to [1, 23, 26]

Stages	Description	Time horizons by selected author		
		Armstrong 1999	Itika 2011	Kaufman 2014
Employee welfare in the industrial age	Collecting and managing employee records	1915–1919	1900–1919	1880–1919
Focus on personnel administration	Administrative services, solving current needs of line management	1920–1939	1920–1939	1920–1929
Personnel management (development phase)	The difference between the terms "governance" and "management"	1940–1959	1940–1959	1930–1959
Personnel management (high level)	Systematic training, workforce planning, more sophisticated selection techniques, rewarding and evaluating employees, focusing on the social climate in the workplace	1960–1979	1960–1979	1960–1969
HRM	Focus on efficiency, some software applications are applied (still not integrated), and the selection process in terms of procedures and methods has been partially standardized	1980–1989	1980–1999	1970–1979
Strategic HRM	Maximizing human potential, culture and performance management, ethics in HR work	Since 1990	Since 2000	Since 1980

revolution is taking place, reflecting the coexistence of the Fourth Industrial Revolution based on technology and the Fifth Industrial Revolution based on societal values and needs, both complementary and mutually reinforcing [52]. These circumstances require attracting and retaining employees who are prepared for the introduction of new DTs in the workplace and see continuous learning as part of the organizational culture. In doing so, not only is the demand for employee competencies changing, but the need to redefine the roles of HR managers as strategic partners of companies is also emerging [7, 47].

Table 3.2 summarizes the key events of each industrial revolution and their impact on HRM development processes, starting from the most recent.

Expert scientists [9, 15] discuss that the transformational journey of industry, supported by modern DTs such as additive manufacturing, autonomous robots, augmented reality, cyber security, dynamic supply chain, etc., will lead to a sixth industrial revolution in the 2050s. It is expected that the culture of people working

Table 3.2 The impact of industrial revolutions on HRM [5, 7, 8, 12, 15, 25, 27, 28, 30, 33, 38, 52]

Duration and nature of industrial revolution	Key events of industrial revolution	Main impacts on HRM
Fifth Industrial Revolution (2021-present): Creativity, personalization of production, collaborative robots, i.e., cobots, nanomaterials, bionics, low-energy production, sustainability, and system integration	•Individualized human–machine interaction technologies •Using AI to discover causality in dynamic systems •Digital twins, i.e., copies of production processes, can be used to predict their progress based on cloud data	•Transition from technological advances to an approach focused on personalizing employee working relationships •Shifting focus from the production value of DTs to the human value of technological innovation
Fourth Industrial Revolution (2011-present): 3D printing, digitalization, cyber-physical systems, cloud storage, IoT, AI, big data, robotic process automation, mobile computing, smart factories	•Google Apps Engine 2008: cloud computing platform •SLS 2006: first commercially available 3D printer •Hadoop 2005: a framework for storing extensive data sets •2002–2004: establishment of social networks LinkedIn, Skype, Facebook •Ashton 1999: creating the IoT •1995: chatbot A.L.I.C.E •Intershop AG 1995: first store for online shopping	•2010–2019: specialized HR portals •2012: 360-degree feedback method in employee evaluation, online training •2000–2009: online recruitment, training systems, and employee performance management systems •Pfizer, Dell, Hewlett-Packard 2000: self-service HR technology for employees •Monster 1994: first online recruitment website
Third Industrial Revolution (1960–1989): Automation of production lines, electronics, and development of IT	•1987–1994: commercialization of the Internet •Google 1988: first service for indexing web pages •PeopleSoft 1987: a purpose-built system for HRM •IBM 1981: computer launch •Morley 1969: first programmable logic automaton •McCarthy and Licklider 1963: the idea of cloud computing introduced •McCarthy 1961: the idea of sharing computer technology	•Ulrich 1997: a multipurpose model for HR departments •Beer and Fombrun 1984: Harvard Model •UK 1976: Employment Non-Discrimination Act on the grounds of nationality and citizenship •UK 1975: Employment Non-Discrimination Act on the grounds of sex and marital status •Becker [5]: human capital theory •India 1960: first external recruitment agency

(continued)

Table 3.2 (continued)

Duration and nature of industrial revolution	Key events of industrial revolution	Main impacts on HRM
Second Industrial Revolution (1870–1959): Large-scale use of electricity, oil, steel, and strip manufacturing	•Samuel 1959: machine learning •1940–1949: first electronic autonomous robot •Ford Motor Company 1913: electrified belt line •Daimler and Benz 1883–1886: petrol engine •Edison 1879: invention of the light bulb •Bell 1876: invention of the telephone •Cincinnati 1870: first assembly line, division of labor	•Moxon [33]: part-time employment, development of factory medical services, canteens, nurseries •General Electric 1950s: first HR information system introduced •First Personnel Department in the mining industry •DuPoint 1921: first organizational structure •US 1913: Institute of Personnel Management is founded •Teta Steel 1912: introduction of the 8-h working day •Taylor 1900s: employee management system •Emergence of unions that advocate for workers' rights
First Industrial Revolution (1760–1869): Mechanization, the transition from manual labor to large-scale machine production, the development of chemistry, transport, engineering, and trade, and industrialization	•US 1869: first transcontinental railway completed •Mors 1837: development of the telegraph •Cartwright 1784: first mechanical loom •Watton 1765: development of the steam engine	•US, UK: the introduction of the HR function for employee record keeping and social care •Increase in the number of employees due to immigrant workers •Smith 1776: recognition of the difference between professions and wages

with advanced manufacturing systems and tools may change, with every operation performed by automated robots being controlled by the human mind.

According to the book's Author, HRM technology will not become utterly autonomous with further automation in the coming decades. On the contrary, HRM's work in promoting a new organizational culture and supporting the development potential of employees in a digitalized work environment will remain indispensable.

3.2 Definition and Essence of Digital Human Resource Management

Digital HRM first came to the forefront of professional discussion half a century ago. A further shift occurred in 1995 [41] when the concept of HRM focused on the strategic idea of DTs [50]. In the last three decades, scholars and practitioners have shown increasing interest in exploring how DT is changing the function of HRM, its design, and processes [34]. With the advent of the Fourth Industrial Revolution in 2011, it is clear that fundamental changes are necessary for both employees and employers. Employees are expected to be ready to learn new competencies and acquire new technological knowledge and socio-emotional skills. Organizations should focus on redesigning their structures, business processes, and organizational culture to reflect new working and management methods [7].

Digital HRM should be distinct from electronic HRM (e-HRM) or the Human Resource Information System (HRIS). This is because e-HRM involves using web-based technologies to provide HR services in an organization that can be accessed by a broader range of stakeholders, i.e., from the HR department to managers to employees [6]. HRIS is one of the tools used in digital HRM to systematically collect, store, maintain, update, and distribute an organization's HR data [45].

The main idea behind digital HRM is to streamline decision-making and HRM functions (e.g., talent acquisition and upgrading employee T&D) through complex analytics, mobile applications, and interactive platforms in real time [18]. It is primarily about adopting DTs per se. However, it is not about anything other than the organization's senior management collaborating with the HR department to select and implement appropriate technologies. It considers HR activities a complex and creative process [40].

Figure 3.1 shows the number of scientific publications in the WoS and Scopus databases whose title, abstract, or keyword includes "digital HRM" (digital HRM, digital human resource management, or digital human resource management).

The search results showed that since 2017, the number of publications related to digital HRM and the popularity of this phenomenon have gradually increased. This has become especially noticeable in the last five years. Between 2019 and 2023, the most significant contributors have been academics and researchers from the US (i.e., 379 and 1,079 WoS and Scopus entries, respectively), China (i.e., 351 and 987 WoS and Scopus entries, respectively), India (i.e., 194 WoS and 503 in Scopus, respectively), the UK (i.e., 193 in WoS and 532 in Scopus, respectively) and Italy (i.e., 175 entries in WoS and 343 in Scopus, respectively).

The authors look at the definition of digital HRM, its application, and its focus from different perspectives. Stephan et al. [43] define digital HRM as integrating SMACIT technologies aimed at automating other areas of HRM for better productivity, redefining how HR processes are delivered, and improving work-life balance in real time. Molotkova et al. [32] identify digital HRM as a flexible approach to employee development, emphasizing actively and effectively leveraging employees' talents and experiences in the face of changes in the digital environment.

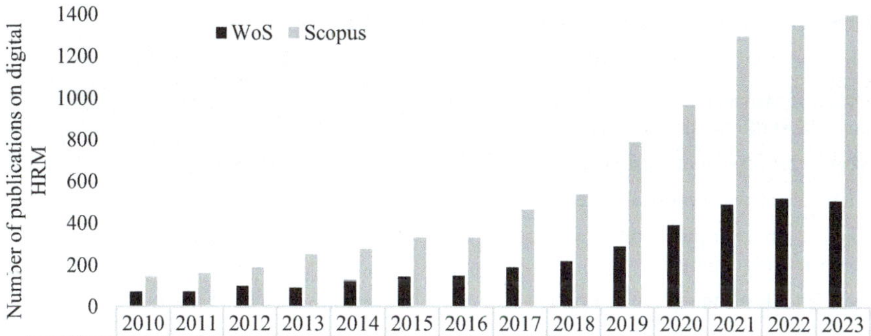

	2010	2011	2012	2013	2014	2015	2016	2017	2018	2019	2020	2021	2022	2023
WoS	72	72	99	91	123	146	151	191	222	292	394	494	523	510
Scopus	148	160	188	252	277	331	332	466	540	792	970	1,297	1,352	1,505

Fig. 3.1 Number of publications on digital HRM in the WoS and Scopus databases

Rana et al. [37] approach digital HRM as an umbrella term encompassing all possible mechanisms of integration between HRM and IT to create internal and external value for target employees and business management.

Strohmeier [44] explains the concept of digital HRM by focusing on its socio-technical outcomes and referring to digital HRM as the process of harnessing the potential of digitalization for operational and strategic HRM purposes. Karaboğa et al. [24] think that digital HRM relies on automating HRM processes and systems in line with the digitalization trend to follow the current HR data.

As stated by [14], digital HRM is the application of advanced technologies and analytics, digital properties, behaviors, and customer orientation from the perspective of the HRM operating model to optimize HR processes to ensure sustainable organizational performance.

Based on the above definitions, digital HRM does not focus on the technical side of HRM. Still, it requires updating the agenda and T&D methods of both HR professionals and the HR organization as a whole. The Author of this book believes that the future of digital HRM lies in building an agile and personalized culture of lifelong learning in the context of the increasing integration of DTs in the workplace to achieve an optimal combination of HRs and digitalization.

Unlike traditional HRM, digital HRM focuses on implementing innovative solutions to increase productivity and views employees as investments that should be supported [32]. As stated by [19], the traditional HRM support model is no longer sufficient to demonstrate the benefits of HRs for organizations. The main features and characteristics of digital HRM compared to traditional HRM are summarized in Table 3.3.

The HR department's role has shifted from providing support and services to employees to being one of the leaders of the organization's advanced digitalization or from a diagnostic to a predictive role. This shift is reflected in implementating competency-based recruitment, rewarding innovation, results-based performance management, and human capital empowerment [39].

Table 3.3 Comparison of traditional HRM and digital HRM [6, 14, 43]

Characteristics of traditional HRM	Characteristics of digital HRM
Database management for administrative support, essential staff records, and periodic reports	Democratized data and advanced HR analytics in real time
Manual processing of personnel documents	Robotic process automation
Hierarchical organizational structure	Agile team networking
Explicit job description with emphasis on specific competencies required	Dynamic competence requirements
Detailed HR planning, job description	Changing nature and typology of work
The focus of recruiters is on self-service	HRs focus on helping employees and increasing their productivity
Thinking focused on improving existing HR processes	Innovative thinking aimed at changing HR processes through DTs

Digital changes in HRM have impacted three key areas (see Fig. 3.2).

Interaction with DTs at work has shaped a new generation of employees or digital HRs who are expected to have a new set of technical and mental skills to acquire and process information systematically. At the same time, DTs have enabled new forms of work organization that range from individual virtual workplaces to virtual groups, teams, or communities to virtual organizations, thus creating the digital workplace. The logical consequence has been the introduction of DTs to support the delivery of

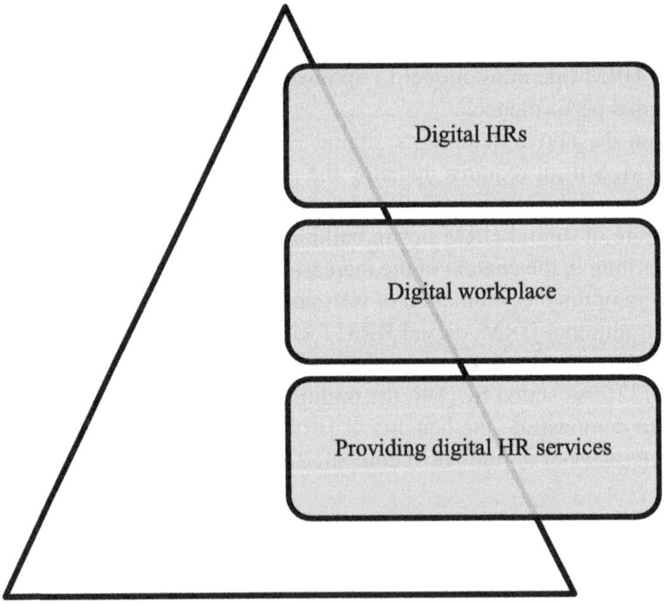

Fig. 3.2 Areas affected by digital HRM [11]

digital services and the implementation of HRM functions, which will ensure lower operational costs, higher speed and quality of HR processes, and the connection between employees, managers, and HR professionals [11].

Based on the above information, research in digital HRM is still in its development phase. New studies are needed to understand the concept better and assess its effectiveness in the socio-technical revolution.

3.3 Trends in the Digitalization of Human Resources

Digitalization has been one of the driving forces in HRM over the last decade, and chief executive officers (CEOs) and HRM leaders worldwide agree. It is highlighted by a survey of 7,096 respondents from 130 countries [43]. More than two-thirds of respondents confirm that digital HRM (74.0%), changing HR skills (78.0%), and the need for employee T&D (84.0%) have reached the forefront of their minds. In terms of industry, the role of digital HRM is significant in the telecommunications (78.0%), financial services (76.0%), and manufacturing (68.0%) sectors.

According to 1,288 HRM leaders from 59 countries, almost one in three employees needs retraining or upskilling. Related to this, the HR department should be a n HR architect, focusing on training leaders to manage in a remote environment and implementing digital learning solutions [29].

Gartner [20] surveyed over 800 HR leaders in 60 countries and identified the top priorities and challenges. These include manager effectiveness, organization design and change management, employee experience, recruitment, and the future of work.

Deloitte was one of the first to focus on a global examination of the priorities, key issues, and accompanying solutions for HRM in response to the crisis caused by the COVID-19 pandemic. The 2019 survey involved 9,456 respondents from 119 countries, two-thirds of whom were in the HRM field [49]. In contrast, the global survey 2021 involved 6,000 respondents, among whom 41.0% were HRM executives.

The comparisons presented in Table 3.4 highlight the changes in approaches to organizational learning, teamwork, and working with HR data influenced by the pandemic.

The impact of the COVID-19 pandemic on the importance, authority, and functional role of the HR department has also been examined by [36] and is outlined in Fig. 3.3.

The need to redefine the organization and establish flexible working arrangements is motivating organizations to consider a hybrid model of working that will help maintain the core values of the business and build a workplace that puts people at the forefront of decision-making.

At the same time, there is a growing demand for employees with transferable competencies who can seamlessly transition into new departments or job roles within the business as needed.

In addition, the HR department is concerned about the security and mental well-being of employees returning to the workplace after working from home.

Table 3.4 Global human capital trends before and during the COVID-19 pandemic, according to Deloitte [13, 49]

HR trends 2019	HR trends 2021
Moving on from the employee experience to a human approach and meaningful work	Enhancing work-life balance through the right choices and setting DTs
Continuing education as a tool to promote employee engagement	Digital learning, increased demands on employees when working with new DTs in terms of resilience, critical thinking, social intelligence, communication and time management skills
Building agile organizations and managing under uncertainty	Combination of DTs and human work as part of teamwork
Increase in the use of DTs in the recruitment process	Redefining HRs work towards developing employee potential and managing change in the context of work automation
Interconnecting HR cloud systems with AI, cognitive technologies, and advanced data analytics	Leveraging real-time data and predictive analytics to transform the HR ecosystem and support employees' physical and mental well-being

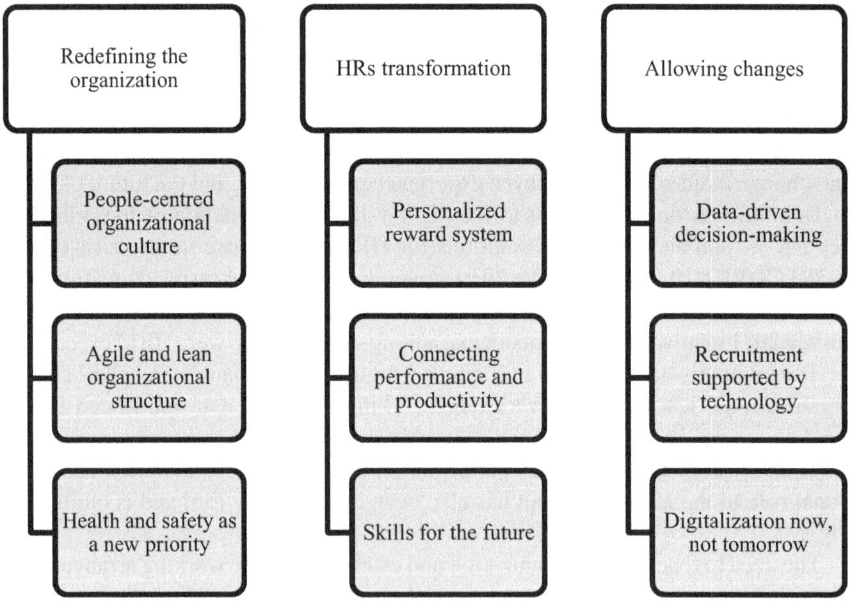

Fig. 3.3 Impact of the COVID-19 pandemic on HRM tasks [36]

Last but not least, the COVID-19 pandemic has also opened up the debate on HR transformation. In the long term, rethinking employee compensation and benefits should aim at personalizing reward structures and performance management based on results. HR departments also need to consider implementing effective training programmes to ensure employees are equipped with the necessary digital competencies and tools in a changing work environment.

Enabling change depends on accepting that digitalization is happening right now. The development and success of remote working have opened the way for alternative recruitment and sourcing strategies, such as virtualizing the entire recruitment process from planning and talent scouting to assessment, selection, recruitment, and onboarding in a virtual environment. DTs such as AI and VR support the HR department's functions, enabling it to push the boundaries of operational efficiency and increase its strategic relevance [36].

Renata Mrazová, head of the HR department at Home Credit, with 118,000 employees, says:

> "Last year, for example, we implemented a new model in which we defined core competencies for all our employees and managers. As a result, principles such as focusing on people or advanced digital competencies have become part of the daily conversation and have gradually been reflected in all key HR processes, which has certainly helped us to better manage the pandemic situation" [13, p. 11].

The above results show that the implications of digitalization for new forms of work and learning are reflected in the organization of the HR department's work and the new requirements for employee competencies.

3.4 Advantages and Disadvantages of Using Technology in Human Resources

Today's organizations are heavily impacted by digitalization in many ways and on many levels, and HRM is no exception. The more HRs use DTs at work, the more the need to link digitalization to companies' overall HRM strategy arises [31]. For digitalization in HRM to be successful, it is necessary to consider all the advantages and disadvantages that the application of DT entails (see Table 3.5).

In the context of the Fourth Industrial Revolution, HR specialists are becoming responsible for adopting new forms of employment and attracting and retaining agile HR adaptable and able to respond quickly to the benefits and risks associated with the Fourth Industrial Revolution [22]. The results of a study conducted among over 2,000 industrial leaders from 26 countries highlight that the biggest challenge brought about by the Fourth Industrial Revolution is the lack of digital culture and employees' digital competencies at work [21].

The research conducted by [40] also suggests that combining creative capabilities and digital competencies represents a focused set of HR characteristics essential for adapting the managerial and organizational model to the impacts of the Fourth

Table 3.5 Key advantages and disadvantages of digitalization in HRM [3, 16]

Advantages of digitalization in HRM	Disadvantages of digitalization in HRM
Better quality with fewer human errors	The staff is reluctant to accept new DTs permanently
Increased operational efficiency	Replacing HRs with automation, computer systems, or robots
Reducing the cost of HRM functions	Cyberattacks or personal data leaks
Acceleration of HRM processes	Changing business model, organizational structure, employee-employer relations
Improving decision-making based on reliable and transparent data	Data integration of separate personnel information systems
Enabling mobility and remote working	Machinery and equipment must respond to HR needs, not vice versa
Improving the overall employee experience	Digital tools still need to be fully used in most HR processes
Strengthening the HR department and taking on the strategic business partner role	Lack of investment in education and development to support employees' digital competencies
Supporting the competitiveness of the company	Slow transformation of HR competencies in response to digitalization

Industrial Revolution. In particular, the optimal combination of creative skills and digital competencies was shown to increase the contribution of ICT to the development of the organizational model for 37.4% of respondents. At the same time, developing procedures and standards that support employees in the transition to a hybrid work model is becoming an unusual challenge for HRM. HR practitioners should be prepared because some employees will need further training to achieve the same efficiency level when working remotely [3].

Other main elements that characterize the perspectives of digitalization in HRM in the following individual HR processes are presented in Table 3.6.

Nabil Tabal, executive director of HRs at the largest oil company with over 65,000 employees, states:

> "As we move deeper into the Fourth Industrial Revolution, we see HR changing to reflect our role as a key business driver. Our success in the future of work will largely depend on our ability to effectively prepare our workforce—by fostering a culture of reskilling, upskilling and lifelong learning" [51, p. 5].

Employees' T&D is a priority within the EU, which has introduced the enhancement of digital skills and competencies for digitalization in the Digital Education Action Plan 2021–2027 [17].

To conclude Chapter 3, the rapid evolution of DTs requires HRs to evolve with them. HRs' current role is to help address concerns about the interaction between people and new digital tools or DTs and to develop a culture and mindset in which this interaction can be successful. There should also be a clear understanding of the

Table 3.6 Challenges created by the Fourth Industrial Revolution in HRM [3]

Processes in HRM	Challenges posed by the Fourth Industrial Revolution in HRM
HR planning	Development of a new planning strategy that addresses the redesign of job profiles and considers the requirements for problem-solving skills, human–machine interaction, and digital competencies
Attracting and recruiting employees	The employee becomes a consumer of the employer's brand and a user of its HRM processes
	There is a need for leaders who can effectively manage change and employees who adapt quickly
	Changing the approach to recruitment to focus more on competencies than on qualifications
Staff T&D	The need to develop new organizational thinking and skills, such as mental cognition, attention control, sense-making, and intuition, to prepare for a dynamic digital work environment
	Providing access to online learning tools and platforms for developing design thinking and digital talent
Employee performance management	The HR department needs more incredible support in overcoming stress, work-life imbalance, and job insecurity
	Efficiency of real-time feedback through the use of DTs and analytical solutions
	Minimizing attrition of highly skilled specialists
Talent retention	Ensure effective communication and opportunities for talent development and upskilling for more complex work and tasks

skills and competencies that employees will need to work in the continuing Fourth and Fifth Industrial Revolutions.

Due to the vastness of HRM, it is not possible in this book to focus on every HR activity affected by digitalization within the primary research. However, the book's Author will, in the following chapters, deal in more detail with identifying and developing competencies required by employers and employees working in selected sectors experiencing an accelerated pace of digitalization.

References

1. Armstrong, M.: A handbook of human resource management practice. Kogan Page, London (1999)
2. Armstrong, M., Taylor, S.: Armstrong's handbook of human resource management practice. Kogan Page, London, New York (2020)
3. Barišič, A., Rybacka Barišič, J., Miloloža, I.: Digital transformation. Enterenova—Ent. Res. In. **7**, 357–366 (2021). https://doi.org/10.54820/GTFN9743
4. Beardwell, I., Holdan, L.: Human resource management: A contemporary perspective. Pitman, London (1994)

5. Becker, G.: Human capital: A theoretical and empirical analysis, with special reference to education. National Bureau of Economic Research, New York (1964)
6. Berber, N., Đorđević, B., Milanović, S.: Electronic human resource management (e-HRM). Strat. Man. **23**, 22–32 (2018)
7. Bissola, R., Imperatori, B.: HRM 4.0: The digital transformation of the HR department. In: Cantoni, F., Mangia, G. (eds.) Human resource management and digitalization, pp. 51–72. Routledge, Abingdon, New York, Torino (2019)
8. Bussler, L., Davis, E.: Information systems: The quiet revolution in human resource management. J. Comp. Inf. Syst. **42**, 17–20 (2002). https://doi.org/10.1080/08874417.2002.11647482
9. Chourasia, S., Tyagi, A., Pandey, S.: Sustainability of Industry 6.0 in global perspective: Benefits and challenges. MAPAN-J. Metr. Soc. of Ind. **37**, 443–452 (2022). https://doi.org/10.1007/s12647-022-00541-w
10. Chruden, H.J., Sherman, A.W.: Managing human resources. South-Western Publishing Company, Nashville (1984)
11. Chytiri, A.P.: Human resource managers' role in the digital era. Spoud.—J. Ec. Bus. **69**, 62–72 (2019)
12. CMCTU: Člověk a stroj: Metodická příručka (Man and machine: methodological guide). Czech-Moravian Confederation of Trade Unions. https://docplayer.cz/68004472-Clovek-a-stroj-kolektiv-autoru-metodicka-prirucka.html (2017). Accessed 6 April 2024
13. Deloitte: Od přežití krize k trvalé prosperitě. Trendy v oblasti lidského kapitálu 2021 v ČR i ve světě (From surviving the crisis to lasting prosperity. Human Capital Trends 2021 in the Czech Republic and the World). https://www2.deloitte.com/content/dam/Deloitte/cz/Documents/human-capital/CZ-HC-trendy-2021.pdf (2021). Accessed 11 June 2024
14. Deloitte: Digital HR: Owning the organization's digital ambitions. https://www2.deloitte.com/dk/da/pages/human-capital/articles/digital-hr.html (2022). Accessed 13 April 2024
15. Duggal, A., Malik, P., Gehlot, A., Singh, R., Gaba, G.S., Masud, M., Alamri, J.F.: A sequential roadmap to Industry 6.0: Exploring future manufacturing trends. IET Comm. **16**, 521–531 (2022). https://doi.org/10.1049/cmu2.12284
16. Erro-Garcés, A.: Industry 4.0: Defining the research agenda. Benchm. **28**, 1858–1882 (2019). https://doi.org/10.1108/BIJ-12-2018-0444
17. EC: Digital education action plan 2021–2027: Resetting education and training for the digital age. SWD (2020) 209 final. https://eur-lex.europa.eu/legal-content/EN/TXT/PDF/?uri=CELEX:52020DC0624&from=EN (2020). Accessed 16 May 2024
18. Galgali, P.: Digital transformation and its impact on organization's human resource management. Rutgers University. https://www.academia.edu/35595152/Digital_Transformation_and_its_Impact_on_Organizations_Human_Resource_Management (2017). Accessed 3 Apr 2024
19. Garcia-Arroyo, J., Osca, A.: Big data contributions to human resource management: A systematic review. The Int. J. Hum. Res. Man. **32**, 4337–4362 (2021). https://doi.org/10.1080/09585192.2019.1674357
20. Gartner: Top 5 priorities for HR leaders in 2023: Actionable and objective advice to tackle top HR challenges. https://www.gartner.com/en/articles/what-will-hr-focus-on-in-2023 (2022). Accessed 17 June 2024
21. Geissbauer, R., Vedso, J., Schrauf, S.: Global Industry 4.0 survey 2016. Industry 4.0: Building the digital industrial enterprise. PwC. https://www.pwc.com/gx/en/industries/industries-4.0/landing-page/industry-4.0-building-your-digital-enterprise-april-2016.pdf (2016) Accessed 3 June 2024
22. Gianecchini, M., Muzzi, C., Campagnolo, D.: HRM 4.0: Digital revolution equals digital competencies? What we expect for workers' competencies in Industry 4.0. In: Cantoni, F., Mangia, G. (eds.) Human resource management and digitalization, pp. 231–242. Routledge, Abingdon, New York, Torino (2019)
23. Itika, J.: Fundamentals of human resource management: Emerging experience from Africa. University of Groningen, Mzumbe University, Leiden, African Studies Centre (2011)

24. Karaboğa, T., Karaboğa, H., Basar, D., Zehir, S.: The digital transformation journey of HR: The effect of big data and artificial intelligence in HR strategies and roles. In: Pejic-Bach, M., Dogru, C. (eds.) Management strategies for sustainability, new knowledge innovation, and personalized products and services, pp. 94–115. IGI Global, Hershey (2022)
25. Kašparová, E.: Digital skills and their development in the Czech Republic. Per. Eng. Nat. Sci. **7**, 637–643 (2019). https://doi.org/10.21533/pen.v7i2.522
26. Kaufman, B.: The historical development of American HRM broadly viewed. Hum. Res. Man. Rev. **24**, 196–218 (2014). https://doi.org/10.1016/j.hrmr.2014.03.003
27. Kohnová, L., Salajová, N.: Industrial revolutions and their impact on managerial practice: Learning from the past. Probl. Pers. Man. **17**, 462–478 (2019). https://doi.org/10.21511/ppm.17(2).2019.36
28. Kolade, O., Owoseni, A.: Employment 4.0: The work of the future and the future of work. Techn. Soc. **71**, 1–15 (2022). https://doi.org/10.2139/ssrn.4073516
29. KPMG: The future of HR in the new reality: Time to start playing the long game. KPMG Lower Gulf Limited. https://assets.kpmg/content/dam/kpmg/es/pdf/2021/02/2020-future-hr-new-reality.pdf (2021). Accessed 15 Apr 2024
30. Lasi, H., Fettke, P., Kemper, H., Feld, T., Hoffmann, M.: Industry 4.0. Bus. Inf. Syst. Eng. **6**, 239–242 (2014). https://doi.org/10.1007/s12599-014-0334-4
31. Mangipudi, M., Vaidya, R.: A study of digitalization in HRM and its effectiveness in execution of HR strategies and policies. Hel. **8**, 4220–4222 (2018). https://doi.org/10.29042/2018-4220-4222
32. Molotkova, N., Makeeva, M., Khazanova, D.: Digitalized personnel management. In: Mantulenko, V. (ed.) The European proceedings of social and behavioural sciences, pp. 757–767. Future Academy, Samara (2019). https://doi.org/10.15405/epsbs.2019.03.75
33. Moxon, G.R.: Functions of a personnel department. Institute of Personnel Management, London (1951)
34. Negi, P., Kaur, J.: The digital workplace: Implications for human resource management. Int. J. Man. Techn. Eng. **9**, 1491–1500 (2019)
35. Obedgiu, V.: Human resource management, historical perspectives, evolution and professional development. J. Man. Dev. **36**, 986–990 (2017). https://doi.org/10.1108/JMD-12-2016-0267
36. PWC: How the new normal is shaping the future of HR? https://www.pwc.com/m1/en/publications/new-normal-shaping-future-hr/how-the-new-normal-shaping-future-hr.pdf (2020). Accessed 16 June 2024
37. Rana, T., Bhakar, S., Kaurav, R., Agraval, V.: Digitalization of human resource practices–an emerging trend. In: Bhakar, S.S., Kaurav, R.P.S., Agraval, V., Gupta, M. (eds.) Proceedings of the 10th international conference of digital strategies for organizational success, pp. 1173–1183 (2019). https://doi.org/10.2139/ssrn.3323740
38. Rotich, K.: History, evolution and development of human resource management: A contemporary perspective. Glob. J. Hum. Res. Man. **3**, 58–73 (2015)
39. Salamzadeh, A., Tajpour, M., Hosseini, E.: Corporate entrepreneurship in University of Tehran: Does human resources management matter? Int. J. Knowl. Bas. Dev. **10**, 276–292 (2019). https://doi.org/10.1504/IJKBD.2019.103218
40. Sergi, B., Klučnikov, A., Popkova, E.: Creative abilities and digital competencies to transitioning to business 4.0. J. Bus. Res. **153**, 401–411 (2022). https://doi.org/10.1016/j.jbusres.2022.08.026
41. Snell, S., Pedigo, P., Kraviec, G.: Managing the impact of information technology on human resource management. In: Ferris, G.R., Rosen, S.D., Barnum, D.T. (eds.) Handbook of human resource management, pp. 159–174. Wiley Blackwell, Cambridge (1995)
42. Stearns, P.: The industrial revolution in world history. Routledge, New York (2012)
43. Stephan, M., Uzawa, S., Volini, E., Walsh, B., Yoshida, R.: Digital HR: revolution, not evolution. In: Bersin, J., Dollar, B., Wakefield, N., Mallon, D., Monck, L., Stempel, J., Vijay, S. (eds.) Global human capital trends 2016. The new organization: different by design, pp. 97–101. Deloitte University Press, New York (2016)

44. Strohmeier, S.: Digital human resource management: A conceptual clarification. Germ. J. Hum. Res. Man. **34**, 345–365 (2020). https://doi.org/10.1177/2397002220921131
45. Tannenbaum, S.: HRIS: User group implications. J. Syst. Man. **41**, 27–32 (1990)
46. Torrington, D., Hall, L.: Personnel management: A new approach. Prentice Hall, London (1991)
47. Ulrich, D., Dulebohn, J.H.: Are we there yet? What's next for HR? Hum. Res. Man. Rev. **25**, 188–204 (2015). https://doi.org/10.1016/j.hrmr.2015.01.004
48. Volini, E., Occean, P., Stephan, M., Walsh, B.: Digital HR: Platforms, people and work. In: Schwartz, V.J., Collins, L., Stockton, H., Wagner, D., Walsh, D. (eds.) Rewriting the rules for the digital age. Deloitte global human capital trends 2017, pp. 87–92. Deloitte University Press, New York (2017)
49. Volini, E., Schwartz, J., Roi, I., Hauptmann, M., Van Durme, Y., Denny, B., Bersin, J.: Deloitte global human capital trends 2019. Leading the social enterprise: Reinvent with a human focus. Deloitte Development LLC. https://www2.deloitte.com/content/dam/insights/us/articles/5136_HC-Trends-2019/DI_HC-Trends-2019.pdf (2019) Accessed 5 May 2024
50. Walker, A.: A brief history of the computer in personnel. Pers. J. **16**, 32–36 (1980)
51. WEF: HR 4.0: Shaping people strategies in the Fourth industrial revolution. World economic forum. https://www3.weforum.org/docs/WEF_NES_Whitepaper_HR4.0.pdf Accessed 4 April 2024 (2019)
52. Xu, X., Lu, Y., Vogel-Heuser, B., Wang, L.: Industry 4.0 and Industry 5.0 - inception, conception and perception. J. Man. Syst. **61**, 530–535 (2021). https://doi.org/10.1016/j.jmsy.2021.10.006
53. Zhong, C., House, J.: Hawthorne revisited: Organizational implications of the physical work environment. Res. Org. Beh. **32**, 3–22 (2012). https://doi.org/10.1016/j.riob.2012.10.004

Chapter 4
Theoretical Background of Competencies and Their Application in the Competency Model

Abstract This Chapter 4 focuses on defining the concept of competency and explaining the essence of competency models, their process of creation, and the approaches applied to it. This Chapter 4 presents the views on the composition of competencies and the requirements for creating competency models represented in the literature. In addition, examples of competency models influenced by digitalization are given. In the next part of the Chap. 4, the importance of digital competencies in business practice in the context of the Fourth Industrial Revolution is emphasized, and selected digital competency models are presented. Due to the limited amount of professional literature available in the Czech language, foreign publications, professional articles, and studies are primarily analyzed here.

4.1 Definition of Competency and Its Components

A wide range of definitions of competency can be found in the literature. The main reason is that the concept has many faces and applications [72] and is defined in foreign literature in two primary meanings, which are "competence" and "competency". It is important to note that some authors [61, 88, 95, 108] distinguish these two meanings according to the area of focus, characteristics, motives, and skilled behaviors, while other authors [62, 68] combine them.

White [106] first used the concept of competence with motivation theories to explain individuals' playful and exploratory behavior. A professor at Harvard University [69] was one of the first to emphasize the importance of competence for individuals to reach their full potential in their personal and professional lives. It has significantly influenced the practice of HRM towards more reliable prediction of individuals' job performance [108]. In contrast, [59] view competence from three perspectives, namely:

- The effective performance of the individual in a social role;
- The individual's ability to adapt;
- The feeling that the individual can influence their environment.

A. Mazurchenko, *The Impact of the Digitalization on the Human Resource Management*, Information Systems Engineering and Management 25,
https://doi.org/10.1007/978-3-031-76790-6_4

In the initial phase of historical development, competences were viewed as individual characteristics necessary to achieve the desired level of employee performance. Over time, the intention and need to manage organizational competences emerged, gradually identifying critical competences to gain competitive advantage in the 1990s [35]. Prahalad and Hamel [83] first introduced the characteristics of organizations' core competences.

Teodorescu [95] refers to competence as a desired end state for individual performance that is a direct pathway to most effectively achieving organizational goals. Sampson and Fytros [88] add that from an HRM perspective, competence is part of the criteria for selecting the most suitable employee for a given task. Hitt et al. [40] show a different point of view, namely that competence is a set of valuable, scarce, and difficult-to -replace resources and skills. According to [16], competence is the sum of an individual's key personal and professional talents, behavioral patterns, and ability to use them. Skorková [92] states that competence can also be seen as the authority and responsibility of an employee to perform certain activities, and external motives influence it. In addition, competence focuses on the individual's approach to the job task when different aspects of previously acquired knowledge and skills are considered [61].

In contrast to the above definitions, which apply a person-centered behavioral approach, competency, in its second meaning, focuses on a task-oriented functional approach [108]. Competency is derived independently of the individual's environment, depends on the individual's behavior under intrinsic motivation, personality traits, and self-concept, and can be enhanced through T&D [82]. Spencer and Spencer [93] were among the first to think of competency as a causal link between an individual's core characteristics and their practical, especially outstanding, performance in a job or situation where they should meet predetermined criteria.

Thanks to [18], the term competency began to be widely used and discussed in management, initially in the US. Since the 1990s, it has become more directly associated with the observable and testable components of competency and their ability to transfer from person to person and from one domain to another [110]. However, it should be noted that many different definitions of competency are presented in the literature for what the main components should include. A total of 12 components of competency have been identified by the Author of this book, a list of which is provided below:

- 1. The knowledge and understanding that an individual has in specific areas;
- 2. Ability or capability, i.e., the characteristics of an individual to perform in some expected way;
- 3. Skill in the sense of the ability to perform an activity;
- 4. Personal practical experience of the individual;
- 5. Values that are important to the individual;
- 6. Motives that bring a desire to act;
- 7. Attitudes that are manifested in an individual's response to specific situations;
- 8. Traits and personal characteristics of an individual;
- 9. An individual's mindset or thought patterns;

- 10. The self-image of the individual, i.e., the perception of oneself;
- 11. An individual behavior;
- 12. The emotions of the individual.

A given component's presence in the competency definition is then indicated in Table 4.1.

Based on the above analysis, for most of the 18 authors, the essential components of competency are individuals' knowledge awareness and ability to perform at work. In addition, half of the authors emphasize that competency involves the personal characteristics of individuals, and a third adds their motivation and attitudes towards specific work situations.

On the other hand, the mentality and way of thinking of individuals, as well as their emotions and personal experiences, are mentioned marginally. In the next part of this book, the Author will draw on the definition proposed by [77], which introduces some new features while responding to the demands of the current labor market:

Table 4.1 Frequency of occurrence of each component in different authors' definitions of competency [5, 13, 29, 30, 42, 56, 64, 68, 72, 73, 77, 80, 82, 87, 93, 97, 102, 109]

Author, year	Competency component											
	1	2	3	4	5	6	7	8	9	10	11	12
National system of occupations 2022 (NSO)	X	X	X		X		X					
Bendassoli et al. 2016	X										X	X
Salleh et al. al. 2015	X		X								X	
Matuska 2012	X			X		X						
Vodák and Kucharčíková 2011	X		X			X		X		X		
Mitchelmore and Rowley 2010	X		X			X		X				
Hroník 2007	X		X	X				X				
Mulder et al. 2006	X	X	X				X					
Draganidis and Mentzas 2006		X									X	
Król 2006	X		X				X					
Dubois et al. 2004	X		X			X		X	X	X		
Athey and Orth 1999	X	X					X				X	
Woodall and Winstanley 1998	X		X		X		X	X				
Mansfield 1996			X					X				
Parry 1996	X		X				X					
Page et al. 1994		X	X					X				
Tucker and Cofsky 1994	X		X		X	X		X				
Spencer and Spencer 1993	X		X		X	X	X	X		X		
Boyatzis 1982	X		X			X		X		X		
Total	16	5	15	2	4	7	7	10	1	4	4	1

> Competencies are the sum of knowledge, skills, abilities, attitudes, and values that enable an individual to apply and develop personally. They are the language used to describe the requirements of a worker in the National Occupational Framework. Competencies express the prerequisites for performing a particular set of activities. They tell what the employee should know, be able to do, and how he or she should behave.

The different components of competencies included in the above definition are defined as follows:

- Knowledge or know-how: an organized set of facts, principles, concepts, ideas and theories, practices, and information acquired over time to support an understanding of a field or topic [15, 26]. It includes human and social knowledge of individuals, and their structured knowledge of organizational rules, processes, and routines [74].
- Skills: refer to the ability to carry out specific processes and apply one's knowledge (both explicit and tacit) to achieve results [26].
- Capability: having enough strength to do something, especially the physical and mental quality to perform activities [96].
- Attitude: expresses an individual's mindset and readiness to act or react to a particular idea, person, or situation in a certain way [26].
- Values: determine what is desirable or morally right for a particular person.

In the following sections, competency will be examined in more detail regarding its integration into competency models.

4.2 The Concept of Competency Models, Their Typology, and Use in Human Resource Management

The complete integration of competencies into organizational life began in the 1980s when they were actively used to identify and predict leadership effectiveness, to align individual behavior with the goals and values of the organization, and with the actions of leaders in the organization [18]. Around that time, [70] suggested us ing the competency model as a decision-making tool that describes the core competencies needed to perform a specific job. This was because the traditional task-based job analysis process no longer provided the flexibility to identify the competencies necessary for success in most jobs.

In the late 1970s, researchers also began to analyze what knowledge, skills, and behaviors HR professionals need to perform their roles successfully [25]. In the 1980s, Ulrich began his study [98], which developed a competency model for HRM practitioners and gradually updated it in 2007, 2012, and 2016. The latest version of this model defines nine core competency domains and 21 competencies. The domains include the skills of the strategic positioner, analyst, creator of critical organizational capabilities, initiator of organizational culture change and other organizational changes, curator of human capital, technology integrator, and responsible for compliance and employee compensation [46].

In the following decade, [23] introduced a holistic professional or vocational competency model emphasizing the intrinsic interconnectedness of cognitive, functional, personal, ethical, and meta-competencies. Based on the holistic approach, [60] developed their competency model that adds social or behavioral competencies to cognitive, functional, and meta-competencies.

In the late 1990s, organizational managers began to apply the core competency framework proposed by [104], which became known as the hierarchical system model.

The scope of applying the competency model began to extend into different areas of HRM. However, the construction of the competency model was based on the fact that it only met the needs of a particular aspect of HRM in a particular part of the organization [58].

Markus et al. [65] characterize the competency model as a list specifying desirable competencies, the structure of which should support the use of competencies across selected HRM functions.

Clempner and Gutierrez [24] emphasize the applicability of competency models as tools for assessing employees in terms of the chosen job role and the required competencies by comparing employee profiles with the ideal.

Sampson and Fytros [88] agree with the previous approach, which is that the competency model minimizes the gap between the expected competencies and the competencies possessed by the employee by designing applicable learning activities.

The authors approach the typology of competence models depending on how they define competency (see Sect. 4.1). According to [57], three basic types of competency models can be encountered:

- Central competency model: all competencies common for all employees, regardless of their position, role, or hierarchical classification.
- Specific competency model: the competencies that make employees successful directly in the company.
- Generic competency model: competencies that are the same for some employees in the organization but are manifested in different behaviors in different types of employees.

Kociánová [54] presents another possible classification aspect, according to which she divides competency models into:

- Universal competency model;
- Managerial competency model;
- Competency model for specific work activities.

It should be noted that a competency model influenced by organizational strategy, culture, and core values must reflect the changes taking place in business processes and, in one form or another, affect various areas of HRM, from recruitment and career planning to T&D. Fig. 4.1 shows this in more detail.

Although "competency model" and "competency framework" are commonly used synonymously, this is untrue. A competency model is a simplified form of any current

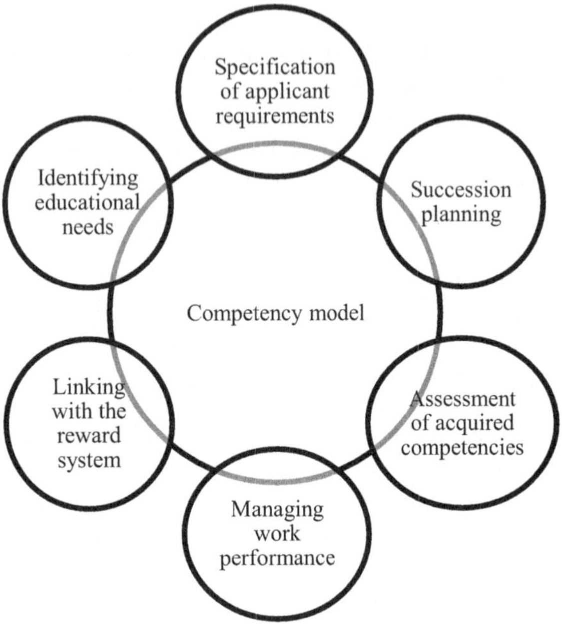

Fig. 4.1 Using the competency model in HRM [79]

or future state of skills, knowledge, attitudes, and behaviors of employees in the workplace.

On the contrary, a competency framework demonstrates the relationship between all aspects of an employee's profile, organizes ideas and makes conceptual distinctions. It can be considered a structure or system for understanding set objectives or results. In other words, a competency framework establishes a competency model and the conditions and time limits for its application and use. In addition, a competency framework can be represented as a tool through which organizations familiarize their employees with the desired behavioral models for specific job roles [49].

4.2.1 Requirements for the Competency Model and the Process of Its Development

Some experts [27, 91, 92, 95] believe that the competency model can only lead to the excellent performance of individuals in a specific work context (i.e., within their work role in a team, department, or entire organization) under the conditions that all its milestones are specific, measurable, observable and clearly defined. From an HRM perspective, it is also crucial that the individual components of the competency model are aligned with the organization's established standards [68].

In addition, several other characteristics are required for a competency model to be functional and applicable to an organization. According to [42], a competency model

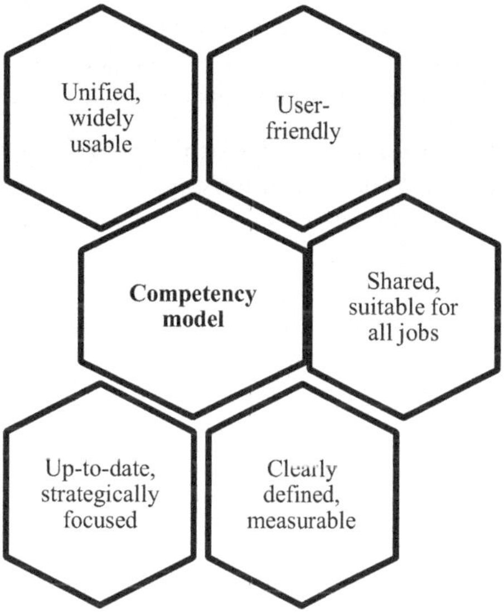

Fig. 4.2 Approaches to creating competency models [42, 91, 92]

should link the HRM strategy to the organization's overall strategy, be easy to use, consistent across all job roles, applicable across all HR activities, and be developed with the participation of all company stakeholders.

Skorková [92] points out the need to clearly define each competency in the competency model, verify its measurability by performance standards and indicators, and regularly update it to reflect the organization's current situation. Figure 4.2 summarizes the competency model's requirements.

The development of a competency model can be approached from different angles. Kubeš et al. [57] state that companies can initially apply an existing universal competency model without changes (i.e., a borrowed approach), then modify the already established competency model for organizational culture, structure, strategies, and other specificities (i.e., a combined approach), and subsequently create a new competency model that requires knowledge not only of the internal organization but also of the external specific conditions of these enterprises (i.e., a tailor-made approach).

Cardy and Selvarajan [22] identify two traditional and two alternative approaches to identifying competencies for further application, given the nature of the competencies and the organizational context in competency models:

- Traditional approach: the job-based approach, in which competencies are derived from an analysis of job requirements, and the future-based approach, as competencies are derived from an analysis of future needs in competencies, given the strategic focus of the business.
- Alternative approach: the value-based approach and the person-based approach, which is based on individuals' personal characteristics rather than on the tasks performed.

Singh Chouhan and Srivastava [91] take a different approach to defining approaches to developing competency models compared to previous authors and present three key concepts:

- The single-job approach: It is the most common approach to modeling competencies developed for single jobs that management and HR consider key for the recruitment and development of employees. Such a model is usually prepared based on customer interviews, observations of selected job holders, and a survey of job applicants. The final model consists of 10–20 core characteristics or skills, each requiring specific behaviors contributing to practical outcomes.
- The one-size-fits-all approach: It is based on defining one set of competencies for a wide range of jobs (e.g., for all managers in a company). The concept of such a model is usually drawn from existing individual job competency models and publications on HR leadership and organizational development.
- A multiple-job approach: It involves creating multiple models for multiple jobs at different levels with different job descriptions and no everyday tasks. In addition, these models should be supported by consistent programmes and tools for performance management, career development, and staff selection for the jobs defined within them. This is a time- and resource-intensive exercise.

Creating a competency model is broken down into the steps shown in Table 4.2.

Armstrong and Taylor [4] argue that competency modeling is based on behavioral analysis, which aims to identify the behavioral dimensions that influence job performance (i.e., personal competencies), and functional analysis to identify technical competencies (i.e., work competencies). Horváthová et al. [41] divide the process

Table 4.2 Stages of the competency model development process [66]

	Competency model development stage
1	• Defining objectives: awareness of the need for the model and its focus, selection of units of analysis, work positions, and timeframe
2	• Ensuring senior management support: obtaining information, resources, and permissions and ensuring staff participation from whom data will be collected
3	• Development and implementation of a communication and training plan: meetings with each stakeholder group through different information channels, reporting on progress by set deadlines
4	• Methodology planning: selecting a representative sample of respondents and using reliable, credible, and acceptable data collection methods
5	• Identifying competencies and creating a competency model: defining competencies from the job description, verifying by professional bodies, revising based on feedback, and giving examples of behavior at different levels of expertise
6	• Application of the competency model: in strategic workforce planning, selection of new employees, T&D of existing employees, performance management, succession planning, and compensation
7	• Evaluating and updating the competency model: evaluating the process of developing it and its value to the organization, looking for ways to improve it as part of a continuous process

of developing a competency model into eight phases, which are mostly designed similarly to [66]. An additional step prior to developing of the competency model in its final form is the clarification of expectations phase, during which the required competency levels for the selected job roles should be agreed upon.

4.2.2 The Role of Digital Competencies in the Competency Model and Their Definition

As the use of DTs expands into all sectors, from business to transport to agriculture, the importance of digital competencies and the need to apply them in the competency models of organizations is increasing. According to [105], 65.0% of the jobs that will employ current students still need to be created, creating new demands on the competency profile of employees. As reported by the [32], in Europe, more than 90.0% of future professional positions require a basic level of digital skills, as well as basic literacy and numeracy skills. Nevertheless, around 42.0% of Europeans need to gain basic digital skills, 37.0% of whom are employees. The picture is similar in Australia, where the government's Digital Economy 2030 strategy predicts that 250,000 new jobs will be created by 2025 through digitalization. In addition, almost 90.0% of the Australian workforce will need digital skills during this period [6].

Work organization and working practices have been substantially transformed by the COVID-19 pandemic, which has contributed to a fundamental change in the working environment, increasing the pressure for collaborative change with employees and supporting them in their T&D. There has been an increased demand for employees' abilities to adapt and learn new things quickly. Basic digital skills, critical thinking, communication skills, and the ability to communicate have also come to the fore, as have collaboration skills, achievement motivation, and self-management, which are highly valued by employers [51]. NSO [76] has also responded to the changes in occupational competency requirements brought about by the pandemic by recognizing proficiency in DTs as one of the future competencies and will progressively define and add digital competencies to job descriptions.

The definition of digital competencies is unique in the literature because the authors disagree on a universal approach to identifying terminology in the competencies needed for successful HRM digitalization. The consequence is that in many publications, the terms "digital literacy" [9, 53], "digital skills" [101], and "digital competencies" [36, 78] have the same meaning, with which the Author of this book disagrees. The reasons for this are as follows:

- Digital literacy: It is often found in macroeconomic and policy studies and encompasses the set of basic skills required for working with digital media, information retrieval, and data processing [53]. Although digital literacy specifies the skills required for working with data, it is not necessarily part of the digital competency framework [67, 107].

- Digital skills: They are part of a comprehensive concept of digital competencies. They are generally more critical than the ability to use apps or operate digital devices to understand and use online content [101]. They also include digital information processing and communication [1], critical thinking skills [94], the ability to work on shared documents and projects in online collaborative platforms [7], and the skills needed to find creative solutions and transfer knowledge to new work settings [8].

The importance of digital competencies has been growing since 2006, when they were specified as one of the eight key competencies for lifelong personal development, active participation in society, employability, and sustainable lifestyles in the EU [19]. Today, the literature offers a variety of views on the definition of digital competencies. [33, p. 15] define "digital competence" as:

> "The confident and critical use of information society technology for work, leisure, and communication. It is based on basic information literacy skills and communication technologies: using computers to acquire, evaluate, store, create, present and exchange information and to communicate and participate in collaborative networks via the Internet".

Calvani et al. [21] believe that digital competencies include flexibly solving technological problems, selecting and evaluating information to create knowledge, and promoting awareness of individual responsibility.

According to [36], digital competencies are a set of knowledge, skills, attitudes, abilities, and awareness that are required when using ICT for task completion, problem-solving, communication, collaboration, effective, critical, independent, flexible, ethical and creative content creation for work, leisure, and socialization.

As [52] argue, digital competencies are associated with the ability to evaluate DT critically, use it meaningfully, and have the technical capacity to manage motivation to engage with digital culture.

Oberländer et al. [78] proposed defining digital competencies at work as core knowledge, skills, abilities, and other characteristics that enable people at work to effectively and successfully perform their tasks related to digital tools and DT.

Grevtseva et al. [39] also note that digital competencies combine technical, information and media, consumer, and communicative competencies.

Bondar et al. [17] established a different approach to identifying digital competencies, which include skills, technologies, knowledge, attitudes, learning theory, the learning process, and the interconnection of all these components.

According to [14], digital competencies are closely related to developing soft skills such as creativity, partnership, and problem-solving. It is considered an approach to effective personal action aligned with the principles of authority and didactic safety.

Digital competencies are multifaceted and can be seen as a tool to support realizing organizational goals, changes in job quality, innovative development, and company competitiveness [43]. Some authors [45, 78] point out that there is often a gap between existing and needed digital HR competencies in practice. This is because each employee has an individual set of competencies, and each job has different requirements for employee competencies.

According to the Author of this book, this needs to be taken into account so that employers can promptly respond to future challenges in the digital workplace.

4.2.3 Competency Model in the Fourth Industrial Revolution

Undoubtedly, HR is the key to the success of digitalization. The emerging technologies brought about by the Fourth Industrial Revolution require individuals to understand the skills and job profiles required to perform complex work tasks and be able to manage their lifelong learning in response [3]. Debate on competencies and qualification development requirements for digitalization is currently at an early stage, with estimates of possible scenarios for developing job requirements and the resulting skill requirements varying widely [90].

Despite this, some authors believe that skills and competencies development will play a central role and that the demands on employees will continue to increase given the rate of further development of the Fourth Industrial Revolution [31, 71]. The next section of this chapter provides examples of competency models that consider the effects of digitalization on employee competency profiles.

Universal Competency Framework (UCF) by CEB Inc. is defined as a three-tiered structure. The first level is represents a set of 112 specific sub-competencies. The second level expresses the relationships between them, mapped to 20 competency dimensions and, in turn, eight generic competency factors [11].

An earlier study by [10], based on factor analyses and multidimensional scaling analyses of critical personal and managerial skills, presented this third level as a model of employee performance in the workplace, named the Great Eight. UCF offers a behavioral approach to modeling competencies as it focuses on the individual and considers competencies that an individual can learn and acquire, as opposed to, for example, personal competencies (see Table 4.3). UCF is a generic basis for creating competency models widely used in corporate practice to describe competency models for specific job roles.

Prifti et al. [84] focused on identifying competencies for jobs requiring higher education to perform effectively and efficiently in the Fourth Industrial Revolution. The authors maintained the structure and relationships between the elements of the Great Eight. They identified 68 competencies relevant for employees with a university degree in Industry 4.0 based on a literature search and four focus groups (FGDs) with 18–25 academics in each FGD. Focus group participants were lecturers with previous experience in companies or higher education teaching and learning in IT, information systems, economics, and engineering. Three of the FGDs included lecturers from different countries in the Europe, Middle East, and Africa (EMEA) region, such as Germany, Austria, the Netherlands, and Egypt. Professors and lecturers from Germany, Austria, and Switzerland attended the last FGD. This competency model is based on a behavioral approach that applies to employees with training in information systems, IT, and engineering and adapts to the needs of the Fourth Industrial

Table 4.3 The second and third levels of UCF [84]

UCF competency framework	
Leadership and decesion-making	Decision-making and initiation of action
	Management and control
Support and cooperation	Collaboration with people
	Aherence to the principles and values
Interaction and presentation	Relationships and networking
	Persuasion and Influencing
	Presentation and sharing information
Analysis amd interpretation	Writing and processing reports
	Using expertise and technologies
	Analysis
Creation and conceptualization	Learning and research
	Creating and innovating
	Formulation of strategies and concepts
Organization and implementation	Planning and organization
	Delivering results and meeting customer expectations
	Compliance with guidelines and procedures
Adapting and coping	Adapting to and responding to change
	Persuasion and influencing
Entrepreneurship and performance	Achieving personal work goals and objectives
	Entrepreneurial and business thinking

Revolution. It reflects the competencies in the competency dimension, "Use of expertise and technology". The authors agree that the main limitations of the model are the limited literature search, the need for further research on the competencies required by the Fourth Industrial Revolution, and the exploration of the possibility of applying the model in practice.

The [12] study focused on transforming the HR competency model in the context of the digital economy transition. The qualitative research was conducted in 2019 in the CR, Greece, Hungary, Poland, Romania, and Russia in more than 700 organizations with more than ten employees. Respondents included employees working at different managerial and organizational levels with decision-making power or consultative authority over the learning and professional development of employees within their organizations. The survey results indicate that only 3.6% of employees have digital skills at the level their employer requires. At the same time, the outcome of this survey is a three-module model of HR competencies that requires differentiated approaches to selecting and developing an organization's HR that is not tied to a specific industry or occupation (see Fig. 4.3).

The authors point out that applying this model requires openness and transparency in the organization. Another limitation is that the company's needs tend to change

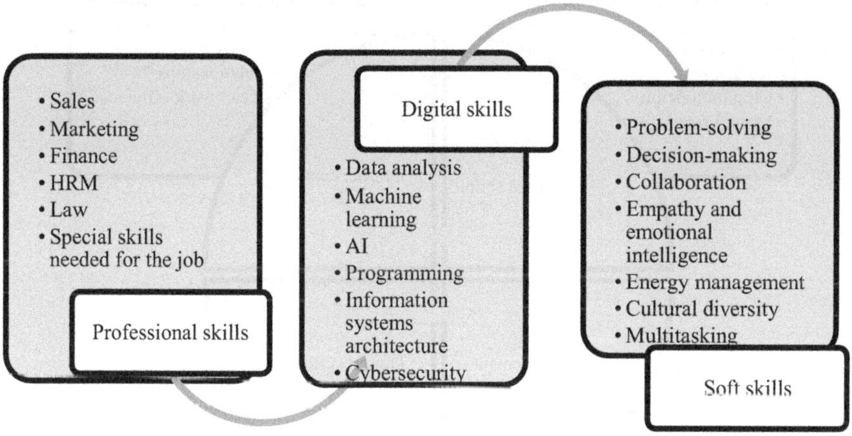

Fig. 4.3 A three-module model of HR competencies in the context of digitalization [12]

in the coming years, and the list of skills within the model should be updated. These skills will be in exceptionally high demand which will also increase the demands on employers for flexible, high-profile professionals.

McKinsey & Company also responded to the need to acquire new skills for the future of work by conducting a global survey of 18,000 respondents from 15 countries, regardless of industry or occupation. The research identified 56 core skills for employees, divided into 13 groups and four categories. One of these categories is digital skills (see Fig. 4.4). The digital literacy group includes digital learning, collaboration, and ethics. The ability to use and develop software implies the mastery of programming literacy, the development of computational and algorithmic thinking, and data and statistical analysis skills. Understanding digital systems includes data and cybersecurity literacy, intelligent systems, and technology support [28].

Competency models can help HR managers support the sustainable development of their organizations in the long term. At the same time, the impact of digitalization on the changing job profiles of HR professionals themselves needs to be considered. In response, [89] used Ulrich's model [99] to guide the development of the HRM Professional Competence models. Based on data collected from 483 South African HRM professionals in 2015, three competency dimensions and 12 associated competencies were identified (see Fig. 4.5). 38.4% (i.e., 184 respondents) had 6–15 years of HRM experience. However, it should be noted that this empirical research on the application of competencies in HRM is limited as it was only conducted in the South African context, and its results cannot be generalized to other countries.

As digitalization gains momentum, HRM professionals must oversee systematic and continuous T&D within the organization and ensure that HR is adapted to the changing labor market conditions.

The Author of this book believes that technological changes and their impact on job profiles should continue to be continuously considered by HR in competency models. This is because the core competencies currently required in most jobs will be

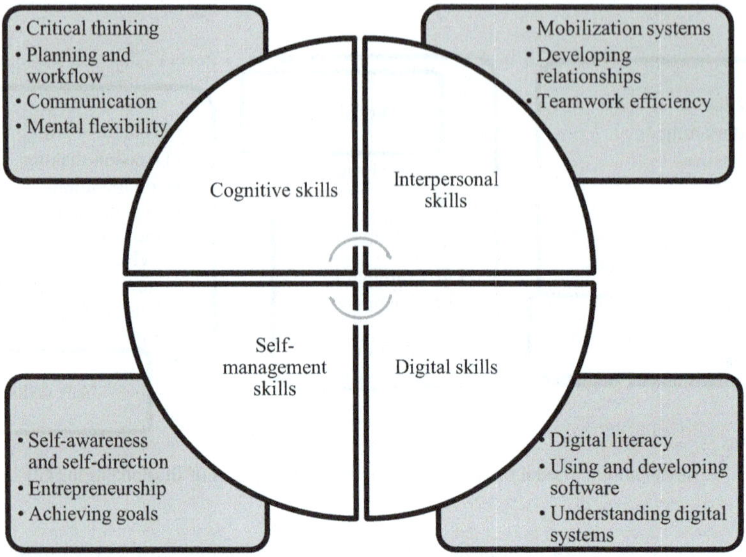

Fig. 4.4 Skills required for the future world of work [28]

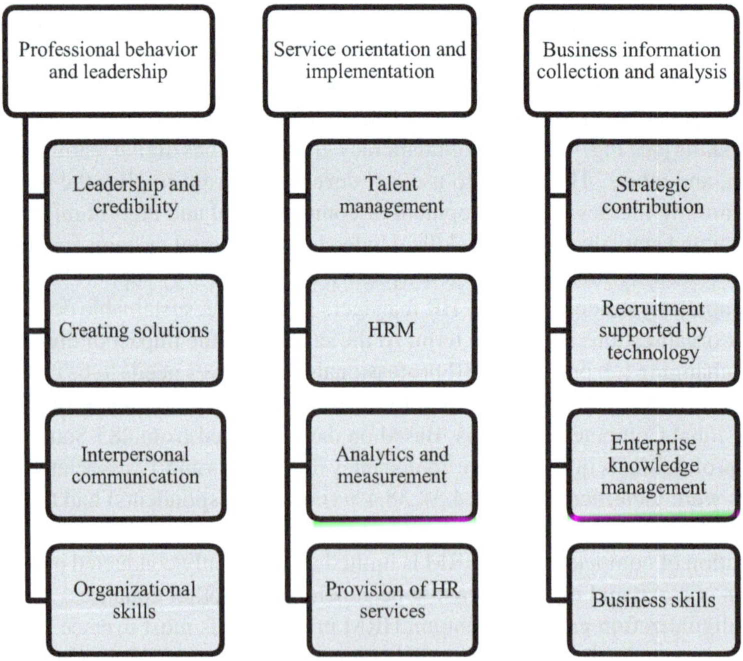

Fig. 4.5 Competency model for HRM professionals [89]

supplemented by additional skills that still need to be considered vital job requirements under the influence of digitalization. It will also affect the work of HRM specialists themselves.

4.2.4 Digital Competency Models

The first global digital competency framework was introduced in 2000 to support the development of national digital strategies in education and training. It was named the International Computer Driving Licence (ICDL) standard. It aimed to enhance digital HR competencies, educators, and society. ICDL certification is available in over 100 countries and has been provided to more than 70 million people worldwide [100]. The ICDL standard measured basic skills (i.e., computer basics, online essentials), the ability to work with office applications (i.e., working with documents, spreadsheets, and presentations, teamwork), and the basics of good IT practices [50]. Two years later, the North Central Learning Lab specified a set of skills important for graduates entering the job market for the first time in 2002. Literacy in the digital age, inventive thinking, effective communication, high productivity, and IT proficiency were identified as key areas [20].

Zobrist and Brandes [113] point out in their survey how the requirements have changed the competencies individuals need in the twenty-first century over the last 15 years. Over time, content, processing, social, complex problem-solving, technical, systems, and resource management skills have come to the fore. Global measurement of adult labor market skills, and in civic and personal life, was conducted for the first time by the OECD in three waves in 39 countries as part of the Programme for the International Assessment of Adult Competencies (PIAAC) in 2011–2017. The survey subjects were basic skills such as literacy, numeracy, and problem-solving [75].

Janssen et al. [45] subsequently provided insight into what a comprehensive digital competency framework looks like. It includes the following areas of digital competencies:

- Hassle free use proving its effectiveness;
- Decisions on appropriate DTs;
- Learning and development through technology;
- Competencies for work and creative expression;
- Use of DTs in everyday life;
- General knowledge and functional skills;
- Legal and ethical aspects;
- Privacy and security;
- Communication and collaboration technologies;
- Understanding the roles of ICT in society;
- A balanced approach to technology;
- Processing and managing information.

The research was conducted among 95 professionals from the education and training, academic research, IT business, and policy sectors. Respondents were from the US, Norway, Israel, Australia, and EU countries, except Poland and Luxembourg. The Delphi method (i.e., online and face-to-face consultations) and concept mapping (i.e., structured conceptualization and visualization) were applied to develop the digital competency framework. The main limitation of this framework is that it gathers expert opinions on what it means to be digitally competent, not facts on the definition of digital competencies [45].

The EC first proposed the Digital Competence Framework for Citizens (DigComp 1.0) in 2013. It aimed to explain how to use and revise digital competencies and identify critical elements of the knowledge, skills, and attitudes needed for digital competencies.

In line with DigiComp 1.0, the 21 digital competencies are grouped into the following five areas: information and data literacy, communication and collaboration, digital content creation, security, and problem-solving [103].

Under the influence of digitalization, DigiComp 2.0 was launched in 2016 with an updated list of the five competency areas mentioned above [111]. According to the Digital Education Action Plan 2021–2027, DigiComp 2.0 should be complemented by including skills related to AI and big data [81]. The next version of DigiComp was released in 2022.

At the same time, the second round of PIAAC is underway to assess the impact of digitalization and the COVID-19 pandemic on individuals' competency profiles. Primary data collection started on 1 September 2022. In the second round, digital competencies are addressed under the adaptive resolution dimension. The final report with the results of the international data collection should be published in the autumn of 2024 [75].

Table 4.4 lists the main occupations perceived to grow in their role and importance in response to accelerated digitalization.

In conclusion, competencies have become prominent in the global labor market because they create value in different sectors and work contexts. Early and accurate identification of personal skills and coordination of employee behavior leads to employee learning and development can be better prepared and implemented at the organizational level.

4.3 The Importance of Training in Organizations for the Development of Employees' Digital Competencies

Organizations must pay attention to regular training to develop digital competencies in employees. As stated by [47], the specifics of each employee's initial training, the application of their knowledge and experience in practice, and the continuity of further learning and development depend on the specific training needs and chosen

Table 4.4 Digital competency frameworks for different professions [2, 37, 38, 44, 48, 55, 86]

Year	Author	Name and occupation	Areas of competence/dimensions
2021	Gartner	Top five digital competencies for finance professionals	Technological literacy; digital translation; digital education; digital bias; digital ambition
2021	Hutchison et al.	Digital engineering competency framework	Data engineering; modeling, and simulation; digital engineering and analysis; systems softwar;, digital enterprise environment
2021	Aggarwal et al.	Digital options for pharmacy staff	Information, data, and content; teaching learning and self-development; communication, collaboration, and participation; technical proficiency; creation, innovation, and research; digital identity, comfort, safety, and security
2020	Jones	Digital competence framework—ICM works—Unlocking the digital future of construction	Digital literacy; security; design and development, analysis and insights on data capture; modeling and simulation; intelligent construction and built assets
2019	KPMG and Skills Impact	Framework for digital skills in agriculture	Digital capabilities (digital literacy; technology operation; data management; data monitoring, analysis, and interpretation; digital communication; incident management) and enabling capabilities (process improvement; personal learning and mastery; collaboration; business transformation; critical thinking)
2017	Government Digital Service Gov.UK	Digital data capability framework and DT for government roles	Specific competencies for data tasks, IT, product and delivery; quality assurance testin; technical and user-centered design work in government
2017	Redecker	Digital competence framework for teachers	Professional engagement, working with digital resources, digital competencies needed for teaching and learning, digital competencies to improve access to assessment, the potential of DT for learner empowerment, and specific pedagogical competencies

training methods. The identification of further training needs is based on an analysis of the current qualification needs of the organization and a view of future needs. Digitalization is now significantly influencing employees' learning and development objectives,emphasizing on acquiring new skills and competencies, learning new ways of working and principles, and changing the means and methods of education [63]. The selection of appropriate training methods and employee development that meet the training programme's objective, meet the organization's requirements , and use

DTs is, therefore, one of the main challenges for modern organizations [102]. Among the essential methods of T&D at work are videoconferencing, video lectures, and e-learning methods, which reduce the time required for learning, change the course of learning and skills acquisition, and increase employee satisfaction with the learning and development process [112].

Research in competencies and their link to organizational learning and development is driven by the pursuit of excellence in performance at the individual and organizational levels. Although identifying competencies is vital for businesses, the competency-based approach is still considered an emerging practice [85]. The Author of this book considers the further development of this approach as inevitable, given the ongoing Fourth and Fifth Industrial Revolutions.

Fejfarová and Fejfar [34] focus their research on finding the relationship between employee T&D and using a competency-based approach. Since 2013, 1,360 organizations of various sizes operating in the CR in different sectors of the economy have participated in the survey. The survey results showed that only one-fifth of the respondents (i.e., 296 organizations) use a competency-based approach. Organizations that use this approach base their assessment (75.7%) and T&D (70.6%) of existing employees on the competency model for the job. However, only 65.2% of organizations using the competency-based approach systematically evaluate the effectiveness of their staff T&D. The results of the Pearson's chi-square test presented by these authors on data from Czech organizations confirmed that organizations that use the competency-based approach place more emphasis on the T&D of their employees than organizations that do not use this approach [34].

Using a competency-based approach to HRM is an effective tool for systematically working with HR within an organization. Based on the above theoretical findings in Chap. 4, applying the competency-based model is a proven approach to investing in HR to achieve higher productivity in the context of the Fourth Industrial Revolution. A well-designed competency model is a valuable way to make recommendations on how to facilitate the development of competencies in each selected job through recruitment, assessment, T&D activities, and programmes.

The literature review for the book is presented in Chaps. 3, 4 and 5 and has certain limitations. First, the choice of keywords influences which publications were included in the literature search, and the term "digitalization" excluded many papers using synonyms. Second, only publications in languages understood by the Author of this book were included. All works in other languages were excluded, even though they may have had different results on the topic under study. The final selection of publications for analysis in the literature section of the book could have been more comprehensive if more than two reputable scholarly databases (i.e., WoS and Scopus) had been used. Given that there is a small amount of research in world literature and research studies on the impact of digitalization accelerated by the COVID-19 pandemic on digital HR competency requirements, this is identified as a gap in knowledge and is explored in more depth in later chapters of this book.

References

1. Aesaert, K., Van Nijlen, D., Vanderlinde, R., Van Braak, J.: Direct measures of digital information processing and communication skills in primary education: Using item response theory for the development and validation of an ICT competence scale. Comp. Ed. **76**, 168–181 (2014). https://doi.org/10.1016/j.compedu.2014.03.013
2. Aggarwal, A., Blochberger, A., Cahill, S., Maisuria, R., Shaikh, A., Wright, P.: Digital capabilities for the pharmacy workforce. Health Education England. https://www.hee.nhs.uk/sites/default/files/Supporting%20Digital%20Literacy%20in%20the%20Pharmacy%20Workforce%20-%20June%202021.pdf (2021). Accessed 16 July 2024
3. Alhloul, A., Kiss, E.: Industry 4.0 as a challenge for the skills and competencies of the labor force: A bibliometric review and a survey. Sci **4**, 34 (2022). https://doi.org/10.3390/sci4030034
4. Armstrong, M., Taylor, S.: Armstrong's handbook of human resource management practice. Kogan Page, London, New York (2020)
5. Athey, T., Orth, M.: Emerging competency methods for the future. Hum. Res. Man. **38**, 215–225 (1999)
6. Australian government: Digital economy strategy 2030. A leading digital economy and society by 2030. Department of the Prime Minister and Cabinet (Australia). https://apo.org.au/sites/default/files/resource-files/2021-05/apo-nid312247.pdf (2021). Accessed 2 June 2024
7. Bălău, N., Utz, S.: Information sharing as strategic behaviour: The role of information display, social motivation and time pressure. Beh. Inf. Techn. **36**, 589–605 (2017). https://doi.org/10.1080/0144929X.2016.1267263
8. Barak, M.: Are digital natives open to change? Examining flexible thinking and resistance to change. Comp. Ed. **121**, 115–123 (2018). https://doi.org/10.1016/j.compedu.2018.01.016
9. Baškarada, S., Koronios, A.: Unicorn data scientist: The rarest of breeds. El. Lib. And Inf. Syst. Prog. **51**, 65–74 (2017). https://doi.org/10.1108/PROG-07-2016-0053
10. Bartram, D.: The great eight competencies: A criterion-centric approach to validation. J. Appl. Psych. **90**, 1185–1203 (2005). https://doi.org/10.1037/0021-9010.90.6.1185
11. Bartram, D.: The SHL universal competency framework. SHL. https://connectingcredentials.org/wp-content/uploads/2015/02/The-SHL-Universal-Competency-Framework.pdf (2012). Accessed 22 June 2024
12. Belolipetskaya, A., Golovina, T., Polyanin, A., Vertakova, Y.: Transformation of the personnel competency model in the context of the transition to the digital economy. In: Zheltenkov, A., Mottaeva, A. (eds.) Proceedings of the E3S web of conferences "Topical problems of green architecture, civil and environmental engineering TPACEE 2019", pp. 1–12. Moscow Region State University, Moscow (2020). https://doi.org/10.1051/e3sconf/202016409005
13. Bendassoli, P., Borges-Andrade, J., Gondim, S., Makhamed, Y.M.: Performance, self-regulation, and competencies of entrepreneurs in Brazilian creative industries. Psic. Teor. E Pesq. **32**, 1–9 (2016). https://doi.org/10.1590/0102-3772e32ne221
14. Bikse, V., Lusena-Ezera, I., Rivza, P.: The development of the digital transformation and relevant competencies for employees in the context of the impact of the Covid-19 pandemic in Latvia. Sust. **13**, 9233 (2021). https://doi.org/10.3390/su13169233
15. Blanchard, N.P., Thacker, J.W.: Effective training: Systems, strategies and practices. Pearson Custom, Boston (2010)
16. Blašková, M., Blaško, R., Kucharčíková, A.: Competences and competence model of university teachers. Proc. Soc. Beh. Sci. **159**, 457–467. https://doi.org/10.1016/j.sbspro.2014.12.407
17. Bondar, I., Komarnitskyi, I., Rusavska, V.: Formation and development of digital skills in university education. Lapl. em Rev. **7**, 47–56 (2021). https://doi.org/10.24115/S2446-6220202017Extra-B882p.47-56
18. Boyatzis, R.E.: The competent manager: A model for effective performance. John Wiley & Sons, New York (1982)

19. Brolpito, A.: Digital skills and competence, and digital and online learning. European training foundation. https://www.etf.europa.eu/sites/default/files/2018-10/DSC%20and%20DOL_0.pdf (2018). Accessed 2 July 2024
20. Burkhardt, G., Monsour, M., Valdez, G., Gunn, C., Dawson, M., Lemke, C., Coughlin, E., Thadani V, Martin C.: Engauge 21st century skills: Digital literacies for a digital age. North Central Regional Educational Laboratory, Metiri Group. https://www.readkong.com/page/engauge-21st-century-skills-literacy-in-the-digital-age-8003382 (2003). Accessed 19 July 2024
21. Calvani, A., Cartelli, A., Fini, A., Ranieri, M.: Models and instruments for assessing digital competence at school. J. E-Learn. Knowl. Soc. **4**, 186–193 (2008). https://doi.org/10.20368/1971-8829/288
22. Cardy, R., Selvarajan, T.: Competencies: Alternative frameworks for competitive advantage. Bus. Hor. **49**, 235–245 (2006). https://doi.org/10.1016/j.bushor.2005.09.004
23. Cheetham, G., Chivers, G.: Towards a holistic model of professional competence. J. Eur. Ind. Train. **20**, 20–30 (1996). https://doi.org/10.1108/03090599610119692
24. Clempner, J., Gutierrez, A.: Prototyping for selecting people: A competencies model. In: Hamza, M.H. (eds.) Proceedings of the 23rd international conference on modelling, identification, and control, pp. 485–494. Acta Press, Calgary (2004)
25. Cohen, D.: HR past, present and future: A call for consistent practices and a focus on competencies. Hum. Res. Man. Rev. **25**, 205–215 (2015). https://doi.org/10.1016/j.hrmr.2015.01.006
26. Council of the EU: Council recommendation of 22 May 2018 on key competences for lifelong learning (text with EEA relevance). Off. J. Eur. Un. **61**, C189/01 (2018)
27. Černý, M.: Models of approaches to digital competences. J. Techn. Inf. Ed. **12**, 47–71 (2020). https://doi.org/10.5507/jtie.2020.019
28. Dondi, M., Klier, J., Panier, F., Schubert, J.: Defending the skills citizens will need in the future world of work. McKinsey & Company Global. https://www.mckinsey.com/industries/public-and-social-sector/our-insights/defining-the-skills-citizens-will-need-in-the-future-world-of-work (2021). Accessed 14 May 2024
29. Draganidis, F., Mentzas, G.: Competency based management: A review of systems and approaches. Inf. Man. Comp. Sec. **14**, 51–64 (2006). https://doi.org/10.1108/09685220610648373
30. Dubois, D.D., Rothwell, W.J., King Stern, D.J., Kemp, L.K.: Competency-based human resource management: Discover a new system for unleashing the productive power of exemplary performers. Davies-Black, Mountain View (2004)
31. Enke, J., Kraft, K., Metternich, J.: Competency-oriented design of learning modules. Proc. CIRP **32**, 7–15 (2015). https://doi.org/10.1016/j.procir.2015.02.211
32. EC: Digital skills. European commission. https://digital-strategy.ec.europa.eu/en/policies/digital-skills (2022). Accessed 12 May 2024
33. European parliament and council of the EU: Recommendation on key competences for lifelong learning 2006/962/EC. Of. J. EU L394, 10–18 (2006)
34. Fejfarová, M., Fejfar, J.: Employee training and development and competency-based approach: Any relationship? J. Eff. Resp. Ed. Sci. **15**, 242–250 (2022). https://doi.org/10.7160/eriesj.2022.150405
35. Fejfarová, M., Urbancová, H.: Application of the competency-based approach in organizations in the Czech Republic. E and M Ek. And Man. **18**, 111–122 (2015). https://doi.org/10.15240/tul/001/2015-1-009
36. Ferrari, A.: Digital competence in practice: An analysis of frameworks. Publication Office of the EU. https://doi.org/10.2791/82116. (2012). Accessed 30 June 2024
37. Gartner: Gartner identifies five digital competencies finance teams will need to reach their goals in 2021. https://www.gartner.com/en/newsroom/press-releases/2021-01-19-gartner-identifies-five-digital-competencies-finance-teams-will-need-to-reach-their-goals-in-2021 (2021). Accessed 12 Apr 2024
38. Government digital service Gov.UK: Digital, data and technology profession capability framework: Roles in government and the skills needed to do them. Central Digital and Data

Office. https://www.gov.uk/government/collections/digital-data-and-technology-profession-capability-framework#contents (2017). Accessed 14 May 2024
39. Grevtseva, G., Bashtanar, I., Isakova, T., Balikaeva, M., Kirillova, Y.: The digital information competence of an individual: Historical aspect. In: Shaposhnikov S. (ed.) Proceedings of the international conference on quality management, transport and information security, information technologies, pp. 474–476. Institute of Electrical and Electronic Engineers, Piscataway (2020)
40. Hitt, M.A., Ireland, R.D., Hoskisson, R.E.: Strategic management: Concepts and cases. Cengage Learning, Boston (2016)
41. Horváthová, P., Bláha, J., Čopíková, A.: Řízení lidských zdrojů. Nové trendy (Human Resource Management. New trends). Management Press, Prague (2016)
42. Hroník, F.: Jak se nespálit podruhé. Strategie a praxe výběrového řízení (How not to get burned a second time.Tender strategy and practice). MotivPress, Brno (2007)
43. Hubschmid-Vierheilig, E., Rohrer, M., Mitsakis, F.: Digital competence revolution and human resource development in the United Kingdom and Switzerland. In: Loon, M., Stewrd, J., Nachmias, S. (eds.) The future of hrd. Volume 1: innovation and technology, pp. 53–91. Palgrave Macmillan, Basingstoke (2020). https://doi.org/10.1007/978-3-030-52410-4_3
44. Hutchison, N., Pepe, K., Blackburn, M., See Tao, H.Y., Verma, D., Whitcomb, W., Khan, R., Peak, R., Baker, A.: WRT 1006 technical report: Developing the digital engineering competency framework (DECF)–phase 2. Technical report SERC-2021-TR-005. Systems engineering research center. https://sercproddata.s3.us-east-2.amazonaws.com/technical_reports/reports/1616668486.A013_SERC%20WRT%201006_Technical%20Report%20SERC-2021-TR-005_FINAL.pdf (2021). Accessed 17 July 2024
45. Janssen, J., Stoyanov, S., Ferrari, A., Punie, Y., Pannekeet, K., Spoel, P.: Experts' views on digital competence: Commonalities and differences. Comp. Ed. **68**, 473–481 (2013). https://doi.org/10.1016/j.compedu.2013.06.008
46. Jesuthasan, R.: HR's new role: Rethinking and enabling digital engagement. Strat. HR Rev. **16**, 60–65 (2017). https://doi.org/10.1108/SHR-01-2017-0009
47. Ježková, P.G.: Reference of methods and goals of further education from the perspective of social workers. Lif. Learn. **11**, 225–241 (2021). https://doi.org/10.11118/lifele20211103225
48. Jones, D.: Digital competencies framework — ICM works — unlocking construction's digital future. The Institute of Construction Management. https://the-icm.co.uk/2020/05/06/digital-competencies-framework-%E2%94%80-icm-works-%E2%94%80-unlocking-constructions-digital-future/ (2020). Accessed 21 July 2024
49. IAEA: The competency framework — a guide for IAEA managers and staff. International atomic energy agency. https://www.iaea.org/sites/default/files/18/03/competency-framework.pdf (2020). Accessed 20 June 2024
50. ICDL Foundation: International computer driving licence for workforce. https://icdl.org/icdl-programmes/ (2015). Accessed 10 July 2024
51. ILO: Global framework on core skills for life and work in the 21st century. Int. Labour Organ. https://www.ilo.org/wcmsp5/groups/public/---ed_emp/---emp_ent/documents/publication/wcms_813222.pdf (2021). Accessed 30 Apr 2024
52. Ilomäki, L., Paavola, S., Lakkala, M., Kantosalo, A.: Digital competence — an emergent boundary concept for policy and educational research. Ed. Inf. Techn. **21**, 655–679 (2016). https://doi.org/10.1007/s10639-014-9346-4
53. Karpati, A.: Digital literacy in education. UNESCO institute for information technologies in education. https://iite.unesco.org/pics/publications/en/files/3214688.pdf (2011). Accessed 22 Apr 2024
54. Kociánová, R.: Personální činnosti a metody personální práce (Personnel activities and methods of personnel work). Grada Publishing, Prague (2010)
55. KPMG, Skills impact: Agricultural workforce digital capability framework. Faethm, University of Queensland. https://www.crdc.com.au/sites/default/files/Agricultural%20workforce%20digital%20capability%20framework_Report_Final%20deliverable.pdf (2019). Accessed 30 June 2024

56. Król, H.: Podstawy koncepcji zarządzania zasobami ludzkimi (Fundamentals of human resource management concepts). In: Król, H., Ludwiczyński, A. (eds.) Zarządzanie zasobami ludzkimi: tworzenie kapitału ludzkiego organizacji (Human resource management: creating an organization's human capital), p. 50–91. Wydawnictvo Naukowe PWN, Warsaw (2006)
57. Kubeš, M., Spillerová, D., Kurnický, R.: Manažerské kompetence (Managerial competencies). Grada Publishing, Prague (2004)
58. Kurz, R., Bartram, D.: Competency and individual performance. In: Robertson, I.T., Callinan, M., Bartram, D. (eds.) Organizational effectiveness, pp. 227–255. John Wiley & Sons, Hoboken (2002). https://doi.org/10.1002/9780470696736.ch10
59. Kuypers, J.A., Bengston, V.L.: Social breakdown and competence. A model of a normal aging. Hum. Dev. **16**, 181–201 (1973). https://doi.org/10.1159/000271275
60. Le Deist, F.D., Winterton, J.: What is competence? Hum. Res. Dev. Int. **8**, 27–46 (2005). https://doi.org/10.1080/1367886042000338227
61. Leung, K., Trevena, L., Waters, D.: Development of a competency framework for evidence-based practice in nursing. Nurs. Ed. Tod. **39**, 189–196 (2016). https://doi.org/10.1016/j.nedt.2016.01.026
62. Lilleväli, U., Täks, M.: Competence models as a tool for conceptualising the systematic process of entrepreneurship competence development. Ed. Res. Int. **2017**, 5160863 (2017). https://doi.org/10.1155/2017/5160863
63. Makarov, S.I., Sevastyanova, S.A.: Information modeling of the students' residual knowledge level. In: Ashmarina, S., Mesquita, A., Vochozka, M. (eds.) Digital transformation of the economy: challenges, trends and new opportunities, pp. 502–509. Springer, New York (2020). https://doi.org/10.1007/978-3-030-11367-4_50
64. Mansfield, R.S.: Building competency models: Approaches for HR professionals. Hum. Res. Man. **35**, 7–18 (1996). https://doi.org/10.1002/(SICI)1099-050X(199621)35:1%3C7::AID-HRM1%3E3.0.CO;2-2
65. Markus, L., Cooper-Thomas, H., Allpress, K.: Confounded by competencies? An evaluation of the evolution and use of competency models. New Zel. J. Psych. **34**, 117–126 (2005)
66. Marrelli, A., Tondora, J., Hoge, M.: Strategies for developing competency models. Adm. Pol. Ment. Health **32**, 533–561 (2005). https://doi.org/10.1007/s10488-005-3264-0
67. Mason, J., Khan, K., Smith, S., Chen, W., Yang, J., Murthys, S.: Literate, numerate, discriminate — realigning 21st century skills. In: Chen, W., Yang, J.C., Murthys, S., Wong, L., Iyer, S. (eds.) Proceedings of the 24th international conference on computers in education. Asia-Pacific Society for Computers in Education, pp. 609–614. Taoyuan, Taiwan (2016)
68. Matuska, E.: Human resource management system based on competences — Scania case study. Hum. Res. Man. Erg. **6**, 130–144 (2012)
69. McClelland, D.C.: Testing for competence rather than for intelligence. Amer. Psych. **28**, 1–14 (1973). https://doi.org/10.1037/h0034092
70. McLagan, P.: Competency models. Train. Dev. J. **34**, 22–26 (1980)
71. Meudt, T., Metternich, J., Abele, E.: Value stream mapping 4.0: A holistic examination of value stream and information logistics in production. CIRP An. **66**, 413–416 (2017). https://doi.org/10.1016/j.cirp.2017.04.005
72. Mitchelmore, S., Rowley, J.: Entrepreneurial competencies: A literature review and development agenda. Int. J. Ent. Beh. Res. **16**, 92–111 (2010). https://doi.org/10.1108/13552551011026995
73. Mulder, M., Weigel, T., Collins, K.: The concept of competence concept in the development of vocational education and training in selected EU member states. A critical analysis. J. Voc. Ed. Train. **59**, 67–88 (2006). https://doi.org/10.1080/13636820601145630
74. Noe, R.A.: Employee training and development. McGraw-Hill Education, New York (2017)
75. NPI CR: PIAAC ve světě (PIAAC in the world). National Pedagogical Institute of the Czech Republic. https://piaac.cz/o-vyzkumu/piaac-ve-svete/ (2022). Accessed 22 Apr 2024
76. NSO: Národní soustava povolání rozšiřuje popis povolání o digitální competence (Národní soustava povolání rozšiřuje popis povolání o digitální kompetence). National Occupational Framework. https://www.nsp.cz/aktuality/narodni-soustava-povolani-rozsiruje-popis-povolani-o-digitalni-kompetence (2020). Accessed 12 Apr 2024

77. NSO: Definice competence (Definition of competence). National system of occupations. https://kompetence.nsp.cz/#:~:text=Kompetence%20p%C5%99edstavuj%C3%AD%20souhrn%20v%C4%9Bdomost%C3%AD%2C%20dovednost%C3%AD,povol%C3%A1n%C3%AD%20popisov%C3%A1ny%20po%C5%BEadavky%20na%20pracovn%C3%ADka (2022). Accessed 11 June 2024
78. Oberländer, M., Beinicke, A., Bipp, T.: Digital competencies: A review of the literature and applications in the workplace. Comp. Ed. **146**, 103752 (2020). https://doi.org/10.1016/j.compedu.2019.103752
79. Pacltová, A.: Co je to kompetenční model a proč je tak důležitý? (Co je to kompetenční model a proč je tak důležitý?) Inspirata Trainings. https://inspirata.cz/2021/11/13/co-je-to-kompetencni-model-a-proc-je-tak-dulezity/ (2021). Accessed 3 April 2024
80. Page, C., Wilson, M., Colb, D.: Management competencies in New Zealand: On the inside, looking in? Ministry of Commerce, Wellington (1994)
81. Paldusová, H.: Dovednosti pro udržitelnou konkurnceschopnost, sociální spravedlnost a odolnost. Podklad k dokumentům č. 9346/20 a 9349/20 (Skills for sustainable competitiveness, social justice and resilience. Background to documents 9346/20 and 9349/20). Chamber of Deputies of the Parliament of the Czech Republic. https://docplayer.cz/199953019-Dovednosti-pro-udrzitelnou-konkurenceschopnost-socialni-spravedlnost-a-odolnost.html (2020). Accessed 15 July 2024
82. Parry, S.B.: Just what is the competency? (And why should you care?). Train. **35**, 58–64 (1996)
83. Prahalad, C.K., Hamel, G.: The core competence of the corporation. Harv. Bus. Rev. **68**, 79–91 (1990)
84. Prifti, L., Knigge, M., Kienegger, H., Krcmar, H.: A competency model for "Industrie 4.0" employees. In: Leimeister, J.M., Brenner, W. (eds.) Proceedings of the 13th international conference on wirtschaftsinformatik, pp. 46–60. Internationalen Tagung Wirtschaftsinformatik, St. Gallen (2017)
85. Quesnay, P., Poumay, M., Gagnaayre, R.: How does French initial osteopathic training value relational competency, patient education, and the competency-based approach? Int. J. Ost. Med. **42**, 43–50 (2021). https://doi.org/10.1016/j.ijosm.2021.10.002
86. Redecker, C.: European framework for the digital competence of educators: DigCompEdu. In: Punie, Y. (ed.) EUR 28775 EN. Publication Office of the EU, Luxembourg. https://doi.org/10.2760/159770 (2017). Accessed 4 June 2024
87. Salleh, K.M., Khalid, N.H., Sulaiman, N.L., Mohamad, M.M., Sern, L.C.: Competency of adult learners in learning: Application of the iceberg competency model. Proc.Soc. Beh. Sci. **204**, 326–334 (2015). https://doi.org/10.1016/j.sbspro.2015.08.160
88. Sampson, D., Fytros, D.: Competence models in technology-enhanced competence-based learning. In: Adelsberger, H., Kinshuk, H., Pawlowski, J., Sampson, D. (eds.) Handbook on information technologies for education and training. International handbook on information systems, pp. 155–177. Springer, Berlin, Heidelberg (2008). https://doi.org/10.1007/978-3-540-74155-8_9
89. Schutte, N., Barkhuizen, N., Van Der Sluis, L.: The validation of a human resource management professional competence model for the South African context. SA J. Ind. Psych. **41**, a1290 (2015). https://doi.org/10.4102/sajip.v41i1.1290
90. Shevyakova, A., Munsh, E., Arystan, M., Petrenko, Y.: Competence development for Industry 4.0: Qualification requirements and solutions. Ins. Into Reg. Dev. **3**, 124–136 (2021). https://doi.org/10.9770/IRD.2021.3.1(7)
91. Singh Chouhan, V., Srivastava, S.: Understanding competencies and competency modelling — a literature survey. IOSR J. Bus. Man. **16**, 14–22 (2014)
92. Skorková, Z.: Competency models in public sector. Proc. Soc. Beh. Scienc. **230**, 226–234 (2016). https://doi.org/10.1016/j.sbspro.2016.09.029
93. Spencer, L.M., Spencer, S.M.: Competence at work. Models for superior performance. John Wiley & Sons, Toronto (1993)

94. Starkey, L.: Evaluating learning in the 21st century: A digital age learning matrix. Techn. Ped. Ed. **20**, 19–39 (2011). https://doi.org/10.1080/1475939X.2011.554021
95. Teodorescu, T.: Competence versus competency. What is the difference? Perf. Impr. **45**, 27–30 (2006). https://doi.org/10.1002/pfi.4930451027
96. Tran, B.: Expatriate selection and retention: Identifying and assessing the other characteristics beyond knowledge, skill, and abilities. In: Camillo, A.A. (ed.) Handbook of research on global hospitality and tourism management, pp. 468–492. IGI Global, Hershey (2015). https://doi.org/10.4018/978-1-4666-8606-9.ch023
97. Tucker, S.A., Cofsky, K.M.: Competency-based pay on a banding platform. ACA J. **3**, 30 (1994)
98. Ulrich, D.: Strategic human resource planning: Why and how? Hum. Res. Plan. **10**, 37–56 (1987)
99. Ulrich, D.: Human resource champions: The next agenda for adding value and delivering results. Harvard Business School Press, Boston (1997)
100. UNESCO: Digital competence frameworks for teachers, learners and citizens. UNESCO international centre for technical and vocational education and training. https://unevoc.unesco.org/home/Digital+Competence+Frameworks/lang=en/id=11#tbar (2022). Accessed 12 Apr 2024
101. Van Laar, E., Van Deursen, A.J.A.M., Van Dijk, J.A.G.M., De Haan, J.: The determinants of 21st-century skills and 21st-century digital skills for workers: A systematic literature review. SAGE Op. **10**, 1–14 (2020). https://doi.org/10.1177/2158244019900176
102. Vodák, J., Kucharčíková, A.: Efektivní vzdělávání zaměstnanců (Efektivní vzdělávání zaměstnanců). Grada Publishing, Prague (2011)
103. Vuorikari, R., Kluzer, S., Punie, Y.: DigComp 2.2. The digital competence framework for citizens with new examples of knowledge, skills and attitudes. Publication Office of the EU. https://doi.org/10.2760/490274 (2012). Accessed 10 July 2024
104. Wang, Y., Wei, J.: A new framework of core competence of corporations: Hierarchy system model. In: Xiaofeng, J. (ed.) Proceedings of the 99th international conference on management science & engineering, pp. 341–345. Yichang (1999)
105. WEF: The future of jobs report 2018. Insight report. https://www3.weforum.org/docs/WEF_Future_of_Jobs_2018.pdf (2018). Accessed 12 May 2024
106. White, R.: Motivation reconsidered: The concept of competence. Psych. Rev. **66**, 279–333 (1959). https://doi.org/10.1037/h0040934
107. Wolff, A., Gooch, D., Cavero Montaner, J.J., Rashid, U., Kortuem, G.: Creating an understanding of data literacy for a data-driven society. J. Com. Inf. **12**, 9–26 (2016). https://doi.org/10.15353/joci.v12i3.3275
108. Wong, S.C.: Competency definitions, development and assessment: A brief review. Int. J. Ac. Res. Progr. Ed. Dev. **9**, 95–114 (2020). https://doi.org/10.6007/IJARPED/v9-i3/8223
109. Woodall, J., Winstanley, D.: Management development: Strategy and practice. Blackwell Business, Oxford (1998)
110. Yuvaraj, R.: Competency mapping — a drive for Indian industries. Int. J. Sci. Eng. Res. **2**, 1–7 (2011)
111. Zhao, Y., Sánchez Gómez, M.C., Pinto Llorente, A.M.: Digital competence in higher education: Students' perceptions and personal factors. Sustainability **13**, 12184 (2021). https://doi.org/10.3390/su132112184
112. Zakharova, A., Bortnik, A., Romanova, M., Romanov, M.N., Nikolaeva, A.D., Kolpakova, A.P.: On unity of systemically important principles for managing the process of development of future bachelors' competences. Espacios **38**, 27–37 (2017)
113. Zobrist, L., Brandes, D.: What key competencies are needed in the digital age? The impact of automation on employees, companies and education. Deloitte. https://www2.deloitte.com/content/dam/Deloitte/ch/Documents/innovation/ch-en-innovation-automation-competencies.pdf (2017). Accessed 22 July 2024

Chapter 5
Selected Secondary Data Related to Digitalization in Human Resource Management and Digital Competencies

Abstract This chapter aims to present a view of the main changes caused by digitalization in the labor market and to specify its impact on the competency profiles of employees based on the secondary data in a global context. Considering the different degrees of digitalization's impact on individual sectors of the economy, those sectors experiencing the most significant changes have been selected and analyzed in more detail. This chapter also examines the impact of digitalization on human resource management in companies and the competencies of human resource specialists, which are becoming critical in the digital age.

5.1 The Impact of Digitalization on the Labor Market and Employee Competencies

The literature review results presented in Chaps. 3 and 4 confirm that digitalization is a significant challenge for employers and employees. This is because the increasing need to use DT at work is changing the competency requirements of employees across occupations and sectors. Digitalization is influencing the nature of work and catalyzing changes in employment patterns, ways of working, and the setting of employee competency profiles, creating the need for a skills revolution. However, the current level of digital competencies among the population needs to be improved, and their development could be faster, which is why digitalization is not progressing as fast as it could [20]. As [13] state, digitalization needs to be understood under these condition to maximize the potential benefits and opportunities for all.

According to a forecast by [47], by 2030, between 75 and 375 million employees will need to develop their competencies through digitalization, automation, and AI and change careers, which could affect 14.0% of the global workforce. Research by [73] also highlights that at least 5.0% of new jobs in the UK in 2030 will be related to robotics, AI, and other DTs that currently do not even exist. According to [45] a survey of nearly 20,000 employers in 42 countries, finding employees with the necessary soft, technical, and digital skills is a global challenge to reduce the risk of replacing human labor with robots and DT. Employees in Australia [28] and the CR

A. Mazurchenko, *The Impact of the Digitalization on the Human Resource Management*, Information Systems Engineering and Management 25,
https://doi.org/10.1007/978-3-031-76790-6_5

[71] believe that digital and technical skills are among the most critical skills for the future of work.

At the same time, exciting insights are provided by a survey conducted by [1], which examined the impacts of digitalization and the challenges faced by small and medium-sized enterprises (SMEs) in France, Germany, Poland, Spain, and the UK. From each of the five countries, 500 businesses participated, with at least a third operating in manufacturing or construction. More than a third of respondents (i.e., 36.0%) indicated that they consider digitalization a top priority, with 40.0% recognizing it as an essential issue. Moreover, on average, almost three out of five SMEs (57.0%) agree that digitalization will positively impact their company's business in the next five years. Respondents perceive concerns about the security of information systems (28.0%) and the need to recruit and retain staff with sufficient digital knowledge and skills (27.0%) as the main barriers to DT adoption. Software development skills (41.0%), mathematics and data skills (35.0%), digital strategy skills (33.0%), and digital project management skills (31.0%) stand out as the most desired skills [1].

At a time when labor markets are struggling to cope with the impact of the COVID-19 pandemic, employers across sectors are unable to find employees with the required mix of technical and soft skills. According to research by [46], which includes a survey of 40,000 employers, at least three in four employers in 40 countries face this challenge. Although this indicator has reached an all-time 16-year high, some sectors are experiencing a more significant shortage of skilled talent than others (see Fig. 5.1).

By 2025, 149 million new digital jobs will be created globally [36], and at least 50.0% of all employees worldwide will need retraining. The demand for employees with advanced digital skills is much higher than the supply of skilled employees, and this gap is increasing. At the same time, the latest Eurostat data from 2021 show that 35.6% of employees in the CR and 27.5% of employees in the EU still have only basic digital skills. Meanwhile, around 3.4% of Czech employees need digital skills at all or digital skills in terms of job performance at a sufficient level. This indicator reached 8.5% in the EU in 2021 [23].

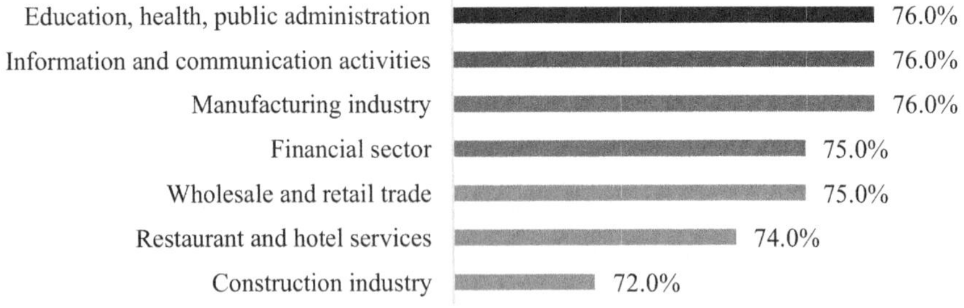

Fig. 5.1 Global talent shortage across sectors [46]

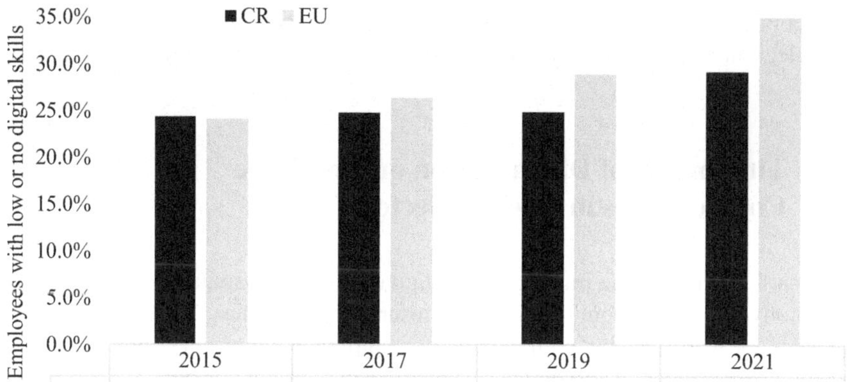

	2015	2017	2019	2021
CR	24.4%	24.8%	24.9%	29.2%
EU	24.1%	26.4%	28.9%	35.1%

Fig. 5.2 Percentage of employees with low or no digital skills in the EU and the CR [23]

Figure 5.2 shows that during the COVID-19 pandemic, the share of employees with low levels of digital skills increased both in the CR and across the EU. which means that organizational T&D is not keeping pace with employers' needs in the face of the accelerated pace of digitalization.

According to data from the DataSkills4SMEs project [22], primarily collected in Italy, Germany, and Austria in 2020, the main barriers to the development of digital HR competencies include the cost of external training agencies (approx. 50.0%) and lack of time due to staff workload (approx. 45.0%). The study includes responses from 338 representatives, mainly SMEs (69.0%), from the manufacturing and service sectors. The self-assessment of digital competencies shows that approximately 40.0% of the respondents reach an intermediate level, and only less than 10.0% rate their level as advanced or very low. Nevertheless, customer relationship management, cloud services, enterprise resource planning systems, and social media are the areas where the need to increase employees' digital competencies is greatest.

Given the relevance and necessity of developing employees' digital competencies globally across occupations and sectors, the Author of this book has reviewed recent studies that have addressed this topic [1, 7, 13, 16, 39, 56]. According to [13], each seventh employer in the UK, Finland, Portugal, Slovakia, Sweden, or Germany identified their employees' digital skills gaps. ECDL Foundation [16] notes that even employees in developed countries such as Switzerland, Denmark, and Austria have significant gaps in digital skills. On the other hand, over 40% of CEOs worldwide regularly update their skills to be effective in the digital environment [39]. One in five employers have found that their employees need further adaptation to new technologies and business processes in the digital age [56]. The findings also suggest that a lack of digital skills or data literacy can be a constrictive barrier to business in the UK, Germany, France, and Poland [1, 7]. These secondary data sources served as a basis for preparing the questionnaire, which was subsequently used in the book to collect primary data in selected sectors of the Czech economy. Section 5.2 justifies the selection of the sector. The survey results are presented in Chap. 6.

5.2 The Impact of Digitalization on Employee Competencies in Selected Sectors

Section 5.2 explores the importance of digitalization in manufacturing (especially automotive), construction, banking, and insurance companies. It includes how DT changes its work and the requirements for employee competence profiles. These sectors were selected based on the secondary data presented below.

Automotive Industry

While the automotive industry in Central Europe has experienced one of the most profound changes in development in recent times [43], its continued success largely depends on, among other things, the response of businesses to the ongoing digitalization accelerated by the coronavirus pandemic. Flexible skills acquisition and continuous T&D are becoming a priority for employees in this sector to succeed in a digitalized work environment [63]. It is vital as 13.8 million Europeans work in the automotive industry. It covers 6.1% of all jobs in the EU and 11.4% of jobs in European manufacturing [2].

It should be mentioned that the CR is among the five European countries with the highest share of jobs in the automotive industry in total manufacturing employment, which equates to 13.7% [2]. With 177,000 people employed in the automotive industry, this sector is key for the Czech economy [57].

By 2030, 16.0% of the global automotive workforce is expected to need retraining and reskilling to meet the needs of organizations driven by digitalization [42]. Even though two out of three companies in the manufacturing industry have provided T&D to increase employees' digital competencies in the last year, there is still a shortage of learning within HR in the European labor market [44]. Due to the rapid development of electric vehicles, there is a growing demand for skills in extensive data collection, analysis, and processing, as well as in data networking, electrical engineering, AI, and robotics [62]. At the same time, advanced analytical and science, technology, engineering, and mathematics skills must not be lacking. However, recruiting employees with such skills is expected to be difficult [54].

In their study, [35] define the DTs most frequently used in automotive companies. These include machine learning (87.0%), big data analytics (84.0%), IoT (82.0%), cloud computing (76.0%), app- and web-enabled markets (76.0%) and autonomous transport (74.0%). Skills gaps and the need for a clear understanding of the opportunities that DT provides (59.0%) are seen as the main barriers to the effective use of DT in the automotive industry. The need for employees with sufficient digital skills to fill positions such as data analyst, computer developer for process automation, AI specialist, and machine learning is expected to continue to increase [35].

Construction Industry

The construction industry is significantly changing due to the Fourth Industrial Revolution. Digitalization can improve productivity and safety at work but should not necessarily reduce overall employment in the construction industry in the long term. In response to them, existing job roles will evolve, and new job roles will be created [14]. By 2025, 31.0% of the job vacancies projected to be offered within European construction companies will be reserved for highly skilled candidates [21], causing the gap between high- and low-level jobs to widen [30]. In the wake of this, promoting the T&D of skilled workers in the construction sector and identifying their skills gaps should be part of the EU policy agenda.

Probst et al. [64] argue that due to ongoing digitalization, only one in three European construction companies has already implemented strategies to retrain their HR. However, one in two respondents from the construction sector stated that they spend at most 5.0% of their annual revenue on employee retraining. Osunsanmi et al. [58] identify other vital areas and barriers to broader digitalization and use of DT in the construction sector:

- Low technical know-how;
- Lack of understanding of how to use new DTs;
- Employees' resistance to change.

As stated by [51], in addition to the automotive industry, the construction industry is another sector where the potential of DTs has not yet been fully explored and exploited. More than a third of European construction companies have difficulties finding employees with the skills required by digitalization [64]. The persistence of a significant shortage of skilled workers with digital skills is also confirmed by construction companies in the CR [19]. It, in turn, reduces the competitiveness of the European construction sector, which provides 14.54 million jobs, or 29.3% of industry jobs. Another study by [32, 33] confirms that construction sector companies understand digitalization's impact on their work better than companies in the manufacturing and automotive sectors (49.0 versus 33.0%). It should also be noted that all sectors struggle with training and recruiting employees with digital skills (24.0%), but this is less relevant for construction than for manufacturing and automotive (13.0 versus 16.0%).

It should be emphasized that views on employees' competencies needed in the construction industry in the context of ongoing digitalization need to be clarified [10, 11, 21, 30, 53]. The authors [30] prioritize ICT security and data processing skills, while the [53] survey results emphasize the importance of innovative learning and practical knowledge management. Other authors also consider flexible thinking [10], digital communication and collaboration [11], accuracy, adaptability, and technical skills [21] strongly required. At the same time, digitalization is expected to require further investment in employee skills that cannot be replaced by robots or DT [65].

Banking and Insurance

Given the growing importance of digitalization, it has had an ambiguous impact on the financial services sector. On the one hand, it helps to increase customer loyalty, improve existing banking and insurance service offerings, reduce human error through technological solutions, and cope with competitors [67]. On the other hand, banks and insurance companies must rethink how they operate, increase the speed of their service delivery, and meet the growing demand for employee competencies needed to use DT effectively at work [24, 26].

A review of recent studies examining the requirements of employees' digital competencies in the banking and insurance sector confirmed that they are pretty different [31, 59, 69, 70]. According to [69], more than half of CEOs in insurance companies working with AI are motivated to learn new skills. Moreover, CEOs in banks agree that VR and augmented reality are helpful in T&D for doing new tasks at work [70]. At the same time, more than 40% of CEOs and employees from the financial sector in different countries rate digital skills as one of the most required skills for senior management [59], considering that only one in seven has advanced digital skills [31].

One of the latest studies by the Economist Intelligence Unit [17], which surveyed 305 banking executives worldwide, revealed the critical DTs needed to transform organizational culture and ensure competitiveness in banking. According to the study, 66.0% of respondents believe that AI, machine learning, blockchain, and IoT will continue to be critical drivers for global banking over the next five years. PWC [66] commented that technical skills, data analytics capabilities, and design thinking are required to successfully digitalize HR in financial services.

According to research provided by [69], only one in four employees in the insurance industry are prepared to work with intelligent technologies. It makes the growing skills gap a significant factor that influences the HR strategy of insurance company managers (43.0%). As pointed out by [32] research, insurance companies are more efficient in understanding the impact of digitalization on their company than companies in other sectors (48.0% vs. 43.0%) but less effective in hiring, T&D of employees with digital skills (19.0% versus 24.0%).

Like construction companies, banks need employees with the skills to understand how DTs can be effectively applied, and they need agile and adaptive HR to navigate these changes [3]. A significant barrier to implementing DT in the banking sector is employee resistance to change and reluctance to learn new things. According to a [70] study, only a quarter of bank employees are ready to work with intelligent technologies, and most cite the growing need for digital skills as a factor affecting bank HRM strategies.

Academic Environment

Continuing digitalization requires education and training systems to adapt to the new context. Current competency requirements have changed significantly, especially in the academic environment. As [38] state, numerous studies have examined the importance of T&D teachers' digital competencies. In confirmation of this, the EC has identified three priority areas for the next few years:

- Better use of DTs for teaching and learning (implementation of the new SELFIE online tool for more effective use of new technologies in training and promotion of digital readiness among teachers, trainers, and students);
- Developing relevant digital competencies for digital transformation;
- Supporting the development of a highly effective digital learning ecosystem through better data analysis and foresight (launching learning analytics and educational AI pilot projects, launching strategic foresight on key trends in digital transformation).

Digital competencies must be among the core skills of all educators for the successful transformation of HEIs and further improvement of education quality and inclusiveness. However, despite efforts to promote the integration of ICT into pedagogical practice, some studies also warn that effective technology adoption by teachers is often not achieved [55]. Consequently, there are plans to develop joint guidelines for teachers and educators to promote digital literacy and combat misinformation through education and training [60].

Different data characteristics play a role in the level of knowledge of digital competencies and respondents. Many studies have compared the academics' digital competencies by field of science, gender, or age.

Table 5.1 provides examples of secondary data related to teachers' digital competencies in higher education settings.

As shown in Table 5.1, several studies have been published in Portugal, Spain, and Ecuador using recent articles on the digital competencies of university teachers. These studies have identified intermediate levels of digital competencies in universities that do not differ much according to gender or age. Most studies also highlight the need for systematic digital competency T&D in universities [29].

The Digital Competence Framework for Teachers shows in a structured and comprehensive way what today's teachers should know to be able to use DT. First, it highlights activities and skills that may be underestimated or not commonly considered in teaching. Second, it primarily focuses on pedagogical competencies and provides insights into how these competencies are affected by the possibility or necessity of using DT in teaching. Finally, teachers do not have digital competencies primarily for themselves but for students and their learning opportunities. Thus, digital competencies are not only about teachers' ability to use DTs but have a broader and more significant impact on their didactic and pedagogical thinking.

As [38] state in their paper, some studies have been published examining digital skills among university faculty and the factors that influence pedagogical adoption of ICT. The most relevant of these are listed in Table 5.2.

Digital skills are also essential in an academic environment to enable teachers to become providers of learning opportunities using the tools, technologies, and resources available at the time. Digitally enhanced teaching and learning have enhanced the examination of teaching techniques and innovative pedagogy and made them more flexible regarding space and time. In particular, professional T&D

Table 5.1 Studies on digital competencies of university teachers [8, 9, 15, 38]

Year	Author	Description/respondents	Key findings
2021	Cabero-Almenara et al.	● The baseline sample was 17,321 professors and 2,180 respondents from Andalusia (Spain) ● Response rate 13.0%	● The average level of digital competencies varies from one industry to another ● Universities need more digital education
2021	Jorge-Vázquez et al.	● The aim is to assess the level of development of digital competencies in acquiring and creating knowledge of a representative sample of university teachers according to the competency framework proposed by UNESCO ● The respondents are 20,466 teachers working in public universities and polytechnics in Ecuador	● College teachers tend to have intermediate levels of digital skills, which are more advanced in generational cohorts that include younger teachers regardless of their gender
2021	Cabral Santos et al.	● The aim is to assess the digital competencies of university professors along several dimensions and analyze how different aspects of these dimensions influence the level of knowledge in these competencies ● The respondents were 35,283 faculty members at Portuguese universities	● Portuguese professors have an intermediate knowledge of digital competencies ● Gender, age group, and career length do not significantly impact professors' level of expertise ● Those who teach Ph.D. students have higher expertise
2020	Dias-Trindade et al.	● The aim is to assess the level of digital competencies of university teachers ● 118 teachers at a Portuguese university	● The level of teachers' digital competencies is medium ● There is a need for teachers to raise their levels of digital skills through specific T&D

is among the three main enablers of digitally enhanced teaching and learning in European HEIs [27].

A recent study conducted in 368 HEIs from 48 countries in the European Higher Education Area [27] shows that 87.0% of teaching staff have a unit in their workplace that supports teachers on all technical issues with which they may not be familiar. In addition, most teachers (87.0%) confirmed that their employers provide training opportunities in digital skills. The findings also show that AI (53.0%), learning analytics (50.0%), machine learning (49.0%), VR (47.0%), and big data (47.0%) are considered the top five strategic development priorities for European HEIs.

Table 5.2 Digital competency frameworks for university teachers [18, 25, 37, 41, 48, 68, 72]

Year	Author	Name	Areas of competency/dimensions/skills
2019	Education and Training Foundation	A professional framework for digital learning: Taking learning to the next level	Curriculum planning, approaches to teaching, supporting students to develop employability skills, subject and sector-specific teaching, assessment, accessibility and inclusion, and self-development
2017	International Society for Technology in Education	ISTE Standards for educators	Using DT to improve student learning, digital leadership and citizenship, collaboration, and facilitating in digital environments, digital learning design, and data analysis
2017	Redecker	European framework for digital competencies for educators (DigCompEdu)	Professional engagement, digital resources, teaching and learning, assessment, student empowerment, facilitating student digital competencies
2017	Kelentrić et al.	Professional digital competencies framework for teachers	Subjects and basic skills, ethics, pedagogy and subject didactics, management of learning processes, interaction and communication, change and development
2012	Fisher et al.	DECK framework	ICT is used for distributed thinking and knowledge, engagement, community communication, and knowledge-building
2011	UNESCO	UNESCO ICT competency framework for teachers	Understanding ICT in education, curriculum and assessment, pedagogy, ICT, organization and administration, and professional teacher education
2006	Mishra and Koehler	Technology pedagogical content knowledge framework	Technological knowledge, pedagogical knowledge, and content knowledge

5.3 The Impact of Digitalization on Human Resource Management and Demands on Employee Competencies

The rapid development of DT brings new functionality to the HRM department and affects how HR activities are performed. Digitalization, robotization, and automation require redesigning the HR role and developing new competencies that will help ensure the prosperity of employees and the organization's sustainability in the digital era [4]. HRM managers who act as strategic partners to the business and are equipped with powerful DTs can drive corporate cultural and technological change and help their company prioritize employees as its most valuable asset [6].

Adopting the new DTs requires more streamlined and standardized processes, systems, and data. Mitrofanova et al. [49] highlight the main challenges associated with it. They include the complexity and heterogeneity of HRM processes and the need to use DT systematically in day-to-day HR activities. The power of digital transformation is only beginning to emerge, and HRM is behind in adopting technology. A survey by Brain and Company [34] confirms that at least one in four businesses still use manual or paper-based processes to deliver HR services. In addition, 75.0% of respondents believe that their digital tools still need to reach their highest level of implementation.

A study of [40] highlights that only one in six companies currently use DT for daily HR T&D. At the same time, most of the tested companies (95.0%) agree that DT brings significant changes and fully or partially supports HR development. According to the research of [61], T&D (51.0%), performance management (45.0%), and onboarding (44.0%) are the areas that organizations want to support digitally in the coming period. HR analytics is also the biggest weakness of the surveyed organizations. A study conducted by [52] reports that big data (15.0%), cloud technologies (10.0%), and mobile technologies (8.0%) are not that popular and are unlikely to be actively used by HR professionals in their daily work in the future. The constant evolution of DT creates new tasks and roles for HRM managers. It motivates them to develop competencies that are shifting from traditional to digital competencies under the influence of digitalization.

In a study by [61], UK HR professionals were asked about their technology readiness, digital mindset, and level of digital skills. The results show that only one in seven respondents identified their team as proficient in various digital skills in social media, mobile apps, analytics, digital learning, and user experience. It highlights that the majority (61.0%) have gaps in digital skills, the largest of which is digital analytics. Another survey [74] also indicates that most HRM professionals (70.0%) need to increase the application of HR technology and analytics skills. At the same time, only 16.0% of HRM leaders feel well prepared to operate in a highly digital environment.

As reported by [12], one in five surveyed HR professionals consider the role of data analyst and digital consultant important for the future of HRM. This is due to the possibility of automating administrative tasks, faster decision-making based on transparent data, and the increasing availability of employee self-service tools. A global study conducted by KPMG [5] summarises the attitudes of HR professionals towards the future of HRM and workplace change in the twenty-first century (see Fig. 5.3). 1201 senior executives in 64 countries, half representing companies with more than 5,000 employees, shared their views in the survey.

Canadian HRM Trends Survey [50], which gathered input from 356 organizations employing more than 680,000 employees, revealed that attracting and retaining employees with the right competencies (59.0%) and helping the organization better adapt to ongoing change (51.0%) are among the top priorities for HR leaders.

Consequently, according to the Author of this book, HRM managers must acquire new competencies to be prepared for current challenges and to provide quick solutions. These include digital employability, the ability to use workforce data to improve business processes, the daily use of DTs to manage changes, the ability to ensure

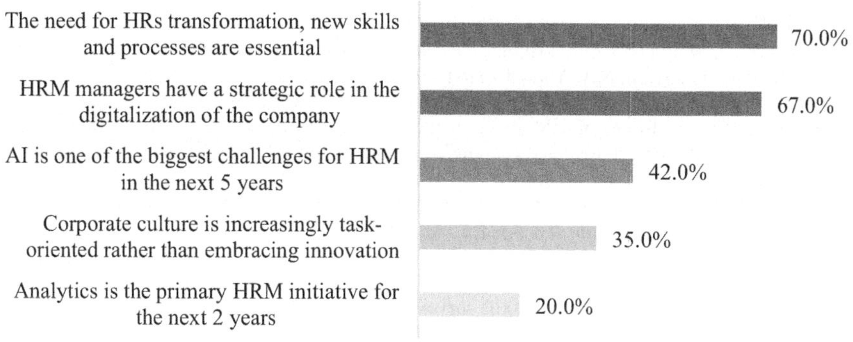

Fig. 5.3 HR managers' views on current key trends in HR [5]

personalized HR services, and the ability to develop new network organizational models to foster an open organizational culture.

From the above findings, it is possible to identify changes in the role of HRM specialists caused by digitalization. HR professionals are expected to be digitally ready to revolutionize the employee experience by integrating people, HR technology, and processes into the new digital ecosystem. This means embracing DT and having the necessary awareness, competencies, and knowledge to leverage it and meet current employee expectations, improve business flexibility, and increase business efficiency.

In conclusion, digitalization demands HR managers to support the employees' T&D in organizations. At the same time, HR managers must have at least basic digital skills to adapt to the new working conditions brought about by the COVID-19 pandemic.

References

1. Abel-Koch, J., Al Abaidi, L., El Kasmi, S., Fernandez Acebedo, M., Morin, L., Tapczewska, A.: Going digital. The challenges facing European SMEs: European SME survey 2019. Bank Gospodarstwa Krajowego, Bpifrance, British Business Bank, Instituto de Crédito Oficial, KfW Bankengruppe. https://www.kfw.de/PDF/Download-Center/Konzernthemen/Research/PDF-Dokumente-Studien-und-Materialien/PDF-Dateien-Paper-and-Proceedings-(EN)/European-SME-Survey-2019.pdf (2019). Accessed 15 Apr 2024
2. ACEA: Share of Direct Automotive Employment in the EU, by Country. European Automobile Manufacturers' Association. https://www.acea.auto/figure/share-of-direct-automotive-employment-in-the-eu-by-country/ (2022). Accessed 12 June 2024
3. Bank Governance Leadership Network, Tapestry Networks, EY: The future of talent in banking: Workforce evolution in the digital era. https://www.tapestrynetworks.com/wp-content/uploads/2024/03/BGLN-ViewPoints-Future-of-talent-in-banking-Final-20-April-2018-web_0.pdf (2018). Accessed 2 Apr 2024

4. Bissola, R., Imperatori, B.: HRM 4.0: The digital transformation of the HR department. In: Cantoni, F., Mangia, G. (eds.) Human Resource Management and Digitalization, pp. 95-113. Routledge, Abingdon, New York, Torino (2019)
5. Bolton, R., Dongrie, V., Saran, C., Ferrier, S., Mukherjee, R., Soderstrom, S., Brisson, S., Adams, N.: The future of HR 2019: In the know or in the no. The gulf between action and inertia. KPMG. https://assets.kpmg/content/dam/kpmg/xx/pdf/2018/11/future-of-hr-survey.pdf (2018). Accessed 11 April 2024
6. Boorman, B., Boroughs, A., Brown, C., Dalton, D., Dinnen, M., Flack, B., Lucas, S., Mes, H., Perring, D., Sonneville, F., Spence, A., Swann, A.: HR trends report 2018: The year of the employee. Oracle. https://www.oracle.com/webfolder/s/delivery_production/docs/FY16h1/doc39/HR-Trends-Report2018-680676.pdf (2018). Accessed 2 May 2024
7. Brain Station: The 2020 digital skill survey: Data survey results. https://brainstation.io/research/digital-skills-survey-2020-results/data (2020). Accessed 1 July 2024
8. Cabero-Almenara, J., Gullién-Gámez, F.D., Ruiz-Palmero, J., Palacios-Rodríguez, A.: Digital competence of higher education professors according to DigCompEdu. Statistical research methods with ANOVA between fields of knowledge in different age ranges. Ed. Inf. Techn. **26**, 4691–4708 (2021). https://doi.org/10.1007/s10639-021-10476-5
9. Cabral Santos, C., Guerreiro, P.N.S., Mattar, J.: Assessment of the proficiency level in digital competences of higher education instructors in Portugal. Ed. (UFSM) **46**, 1–37 (2021). https://doi.org/10.5902/1984644461414
10. CITB: Unlocking construction's digital future: A skills plan for industry. Construction Industry Training Board. https://www.citb.co.uk/media/0pkin1nj/citb_constructions_digital_future_report_oct2018.pdf (2018). Accessed 17 May 2024
11. CLC: Future skills report. Construction Leadership Council. https://www.constructionleadershipcouncil.co.uk/wp-content/uploads/2019/06/CLC-Skills-Workstream_Future-Skills-Report_June-2019_A4-Print-Version.pdf (2019). Accessed 24 June 2024
12. Crummenerl, C., Jacob, F., Funk, L.: Now or never—HR's need to shape its own future. Results and insights from Capgemini's exploration "The future role of HR". https://www.capgemini.com/consulting-de/wp-content/uploads/sites/32/2018/05/now-or-never-hrs-need-to-shape-capgemini-consulting.pdf (2018). Accessed 25 Mar 2024
13. Curtarelli, M., Gualtieri, I.N., Shater, M., Janatti, Donlevy, V.: ICT for work: Digital skills in the workplace. Publication Office of the EU (2016). https://doi.org/10.2759/498467Accessed2July2024
14. De Soto, B., Agustí-Juan, I., Joss, S.: Implications of Construction 4.0 to the workforce and organisational structures. Int. J. Con. Man. **22**, 205–217 (2019). https://doi.org/10.1080/15623599.2019.1616414
15. Dias-Trindade, S., Moriera, J.A., Gomes Ferreira, A.: Assessment of university teachers on their digital competences. QWER. **15**, 50–69 (2020). https://doi.org/10.30557/QW000025
16. ECDL Foundation: Perception and reality: Measuring digital skills in Europe, India and Singapore. https://icdl.org/perception-reality-measuring-digital-skills-gaps-in-europe-india-and-singapore/ (2016). Accessed 1 July 2024
17. Economist Intelligence Unit: Forging new frontiers: Advanced technologies will revolutionise banking. https://www.temenos.com/wp-content/uploads/2020/05/EIU-2020-report-2020-May-27.pdf (2020). Accessed 18 Apr 2024
18. Education and Training Foundation: Taking learning to the next level: Digital teaching professional framework. Full reference guide. https://www.et-foundation.co.uk/wp-content/uploads/2018/11/181101-RGB-Spreads-ETF-Digital-Teaching-Professional-Framework-Full-v2.pdf (2019). Accessed 1 Apr 2024
19. EC: Blueprint for sectoral cooperation on skills. Responding to skills mismatches at sectoral level. A key action of the new skills agenda for Europe. Publications Office of the EU. https://op.europa.eu/en/publication-detail/-/publication/1593ddc1-fcb9-11e6-8a35-01aa75ed71a1/language-en (2017). Accessed 30 June 2024
20. EC: Index digitální ekonomiky a společnosti (DESI). Zpráva z roku 2019. Česko (Digital Economy and Society Index (DESI). 2019 Report. Czech Republic). https://ec.europa.eu/newsroom/dae/document.cfm?doc_id=59981 (2019). Accessed 8 July 2024

21. European Construction Sector Observatory: Improving the human capital basis. Analytical Report. EC. https://ec.europa.eu/docsroom/documents/41261 (2020) Accessed 9 Apr 2024
22. European Entrepreneurs CEA-PME: DataSkills4SMEs: All data skills survey results. https://www.dataskills4smes.eu/2020/11/03/all-dataskills-survey-results/ (2020). Accessed 7 July 2024
23. Eurostat: Individuals' level of digital skills. Statistical office of the EU. https://ec.europa.eu/eurostat/databrowser/view/isoc_sk_dskl_i21/default/table?lang=en (2022). Accessed 14 July 2024
24. Evdokimova, Y., Bondarenko, A., Shinkareva, O.: Digital banks: Development trends. In: Animitsa, E., Dvoryadkina, E., Blaginin, V. (eds.) Proceedings of the 2nd International Scientific Conference on New Industrialization: Global, National, Regional Dimension (SICNI 2018), pp. 151–153. Atlantis Press, Paris (2018). https://doi.org/10.2991/sicni-18.2019.30
25. Fisher, T., Denning, T., Higgins, C., Loveless, A.: Teachers' knowing how to use technology: Exploring a conceptual framework for purposeful learning activity. Curr. J. **23**, 307–325 (2012). https://doi.org/10.1080/09585176.2012.703492
26. Folea, A., Pejkovska Kaeva, M.: Analysis of the digital skills in the EU labor market. A case study of the banking sector. Eur. J. Econ. Bus. Stud. **5**, 66–78 (2019). https://doi.org/10.26417/ejes.v5i2.p66-78
27. Gaebel, M., Zhang, T., Stoeber, H.: Digitally enhanced learning and teaching in European higher education institutions. European University Association. https://eua.eu/downloads/publications/digihe%20new%20version.pdf (2021). Accessed 3 May 2024
28. Gallagher, S.: Peak human potential: Preparing Australia's workforce for the digital future. Centre for the New Workforce Swinburne University of Technology (Australia). https://www.swinburne.edu.au/media/swinburneeduau/centre-for-the-new-workforce/cnew-national-survey-report.pdf (2019). Accessed 21 June 2024
29. Garzón-Artacho, E., Sola Martínez, T., Romero-Rodríguez, J.M., Gómez-García, G.: Teachers' perceptions of digital competence at the lifelong learning stage. Hel. **7**, 1–8 (2021). https://doi.org/10.1016/j.heliyon.2021.e07513
30. Grosso Sategna, L., Meinero, D., Volonta, M.: Digitalising the construction sector. Unlocking the potential of data with a value chain approach. Committee for European Construction Equipment, Innovation and Development Consulting, S.I.R.I.O. https://issuu.com/cece_europe/docs/final_nobianche2-ied_consulting_-_f (2019). Accessed 10 May 2024
31. Hach, W., Stegner, S., Beckert, R.: The digitization race: Can financial service providers hack the pace? Roland Berger. https://www.rolandberger.com/publications/publication_pdf/roland_berger_retail_banking_survey_en.pdf (2018). Accessed 17 Mar 2024
32. Harvey Nash, KPMG CIO: The transformation CIO. Harvey Nash/KPMG CIO survey 2018 construction/engineering industry findings. https://assets.kpmg/content/dam/kpmg/xx/pdf/2018/09/cio-survey-2018-construction-engineering-sector.pdf (2018a). Accessed 9 July 2024
33. Harvey Nash, KPMG CIO: The transformation CIO. Harvey Nash/KPMG CIO survey 2018. Manufacturing/automotive industry findings. https://assets.kpmg.com/content/dam/kpmg/se/pdf/komm/2018/CIO-Survey-2018-manufacturing-auto-report-sweden.pdf (2018b). Accessed 9 Apr 2024
34. Heric, M.: HR's new digital mandate. Digital technologies have become essential for HR to engage top talent and add value to the business. Brain and Company. https://www.bain.com/contentassets/3dea09cc27fd426abfb35f9caa0e97dc/bain_brief_-hrs_new_digital_mandate.pdf (2018). Accessed 20 Mar 2024
35. Hernandez-De-Menendez, M., Morales-Mendez, R., Escobar, C.A., MCgovern, M.: Competencies for Industry 4.0. Int. J. Int. Des. Man. **14**, 1511–1524 (2020). https://doi.org/10.1007/s12008-020-00716-2
36. Hindle, S., Muller, G.: Digital skills AOTEAROA. Digital skills for our digital future. New Zealand Digital Skills Forum, NZTech. https://nztech.org.nz/wp-content/uploads/sites/8/2021/01/Digital-Skills-Aotearoa-Report-2021_online.pdf (2020). Accessed 18 Apr 2024

37. International Society for Technology in Education: ISTE standards for educators. https://www.iste.org/standards/iste-standards-for-teachers (2017). Accessed 21 Mar 2024
38. Jorge-Vázquez, J., Náñez Alonso, S., Fierro Saltos, W., Pacheco Mendoza, S.: Assessment of digital competencies of university faculty and their conditioning factors: A case study in a technological adoption context. Ed. Sci. **11**, 637 (2021). https://doi.org/10.3390/educsci11100637
39. Kane, G.C., Palmer, D., Phillips, A.N., Kiron, D., Buckley, N.: Coming of age digitally: Learning, leadership, and legacy. MIT Sloan Management Review, Deloitte University Press. https://www2.deloitte.com/us/en/insights/focus/digital-maturity/coming-of-age-digitally-learning-leadership-legacy.html (2018). Accessed 10 Apr 2024
40. Katolik Kovačević, A., Cobović, M.: Impact of digital technology in the development of human resources. In: Bacher, U., Barković, D., Dernoscheg, K.H., Maronić, M.L., Matić, B., Pap, N., Runzheimer, B. (eds.) Proceedings of the 13th Interdisciplinary Management Research Conference, pp. 312–326. Josip Juraj Strossmayer University of Osijek (2017)
41. Kelentrić, M., Helland, K., Arstorp, A. T.: Professional digital competence framework for teachers. Norwegian Centre for ICT in Education. https://www.udir.no/contentassets/081d3aef2e4747b096387aba163691e4/pfdk-framework.pdf (2017). Accessed 6 Mar 2024
42. Knoedler, D., Wollschlaeger, D., Stanley, B.: Automotive 2030: Racing towards a digital future underway. IBM. https://www.ibm.com/downloads/cas/NWDQPK5B (2020). Accessed 29 June 2024
43. Kurekova, L.: The automotive industry in Central Europe: A success? IZA Wor. Of Lab. **448**, 1–10 (2018). https://doi.org/10.15185/izawol.448
44. Make UK, SAGE: Digital skills for a digital manufacturing future underway. Manufacturers' Ogranisation Make UK. https://www.makeuk.org/-/media/news-press-release-blog-content/digital-skills/make-uk-sage-digital-skills-for-a-digial-manufacturing-future---final.p (2020). Accessed at 15 July 2024
45. ManpowerGroup: Skills revolution 2.0 robots need not apply: Human solutions for the skills revolution. https://www.skillsforemployment.org/sites/default/files/2024-01/edmsp1_211026.pdf (2018). Accessed 3 July 2024
46. ManpowerGroup: The 2022 global talent shortage. https://go.manpowergroup.com/hubfs/Talent%20Shortage%202022/MPG-Talent-Shortage-Infographic-2022.pdf (2022). Accessed 15 June 2024
47. Manyika, J., Lund, S., Chui, M.: Jobs lost, jobs gained: Workforce transitions in a time of automation. McKinsey Global Institute. https://www.mckinsey.com/~/media/mckinsey/industries/public%20and%20social%20sector/our%20insights/what%20the%20future%20of%20work%20will%20mean%20for%20jobs%20skills%20and%20wages/mgi%20jobs%20lost-jobs%20gained_report_december%202017.pdf (2017). Accessed 28 June 2024
48. Mishra, P., Koeher, M.: Technological pedagogical content knowledge: A framework for teacher knowledge. Teac. Col. Rec. **108**, 1017–1054 (2006). https://doi.org/10.1111/j.1467-9620.2006.00684.x
49. Mitrofanova, E.A., Konovalova, V.G., Mitrofanova, A.E.: Opportunities, problems and limitations of digital transformation of HR management. In: Mantulenko, V. (ed.) The European proceedings of social and behavioural sciences, pp. 1717–1727. Future Academy, Samara (2019). https://doi.org/10.15405/epsbs.2019.03.174
50. Morneau Shepell: Human resources trends for 2019. Insights on what HR leaders are expecting in the coming year: Summary report. https://www.morneaushepell.com/permafiles/91423/hr-trends-2019-summary-report.pdf (2018). Accessed 13 Mar 2024
51. Muñoz-La Rivera, F., Mora-Serrano, J., Valero, I., Oñate, E.: Methodological-technological framework for Construction 4.0. Arch. Comp. Met. Eng. **28**, 689–711 (2021). https://doi.org/10.1007/s11831-020-09455-9
52. Nankervis, A., Connell, J., Cameron, R., Montague, A., Prikshat, V.: Are we there yet? Australian HR professionals and the Forth Industrial Revolution. As. Pac. J. Hum. Res. **59**, 3–19 (2019). https://doi.org/10.1111/1744-7941.12245

53. Oliver Wyman: Digitalisation of the construction industry: The revolution is underway. https://www.oliverwyman.com/our-expertise/insights/2018/sep/digitalization-of-the-construction-industry.html (2018a). Accessed 5 June 2024
54. Oliver Wyman: Building the automotive workforce for the future. Tomorrow's workforce will be lean and digitally adept. https://www.oliverwyman.com/our-expertise/insights/2018/sep/automotive-manager-2018/resources/building-the-automotive-workforce-for-the-future.html (2018b). Accessed 5 June 2024
55. Olofsson, A., Lindberg, J., Fransson, G.: What do upper secondary school teachers want to know from research on the use of ICT and how does this inform a research design? Ed. Inf. Technol. **22**, 2897–2914 (2017). https://doi.org/10.1007/s10639-017-9590-5
56. O'Loughlin, M., Pareek, S., O'Connor, B., Duffield, G., Glynn, N.: CCC global digital skill survey 2019. Cloud Credential Council. https://www.cloudcredential.org/blog/the-ccc-global-digital-skills-survey-2019-report-is-out/ (2019). Accessed 17 May 2024
57. Ortová, M.: Průmysl tenkrát a dnes (Industry then and now). Stat. Us **10**, 1 (2020)
58. Osunsanmi, T.O., Aigbavboa, C.O., Oke, A.E., Liphadzi, M.: Appraisal of stakeholders' willingness to adopt Construction 4.0 technologies for construction projects. Bu. Environ. Proj. As. Manage. **10**, 547–565 (2020). https://doi.org/10.1108/BEPAM-12-2018-0159
59. Oxford Economics, SAP Success Factors: Digital leadership in financial services. https://www.oxfordeconomics.com/publication/open/272816 (2017). Accessed 17 Mar 2024
60. Paldusová, H.: Dovednosti pro udržitelnou konkurnceschopnost, sociální spravedlnost a odolnost. Podklad k dokumentům č. 9346/20 a 9349/20 (Skills for sustainable competitiveness, social justice and resilience. Background to documents 9346/20 and 9349/20). Chamber of Deputies of the Parliament of the Czech Republic. https://docplayer.cz/199953019-Dovednosti-pro-udrzitelnou-konkurenceschopnost-socialni-spravedlnost-a-odolnost.html (2020). Accessed 15 July 2024
61. Patmore, B., Somers, J., D'souza, D., Welch, D., Lawrence, J.: Research report: The state of digital HR in 2017. HRZone. https://www.hrzone.com/resources/the-state-of-digital-hr-in-2017 (2017). Accessed 3 June 2024
62. Pavlásek, P.: Competency-based automotive/mobility engineering education: Implementation of creative project in modular model. In: Proceedings of the Conference ELEKTRO 2020, pp. 1–6. Institute of Electrical and Electronics Engineers, Piscataway (2020). https://doi.org/10.1109/ELEKTRO49696.2020.9130344
63. Plumanns, L., Janssen, D., Vossen, R., Isenhardt, I.: Organisational and individual factors for training of the manufacturing workforce in digitalisation. In: Proceedings of the 2019 IEEE Global Engineering Education Conference (EDUCON), pp. 1158–1166. Institute of Electrical and Electronics Engineers, Piscataway (2019). https://doi.org/10.1109/EDUCON.2019.8725042
64. Probst, L., Lefebvre, V., Martinez-Diaz, C., Unlu Bohn, N., Kitou, D., Conrads, J.: Digital transformation scoreboard 2018. EU businesses go digital: Opportunities, outcomes and uptake. Publications Office of the EC. https://doi.org/10.2826/821639 (2018). Accessed 4 July 2024
65. Prokhin, E.: Digital transformation of industrial companies: What is management 4.0? In: Proceedings of the 2020 11th International Conference on E-Business, Management and Economics ICEME 2020, pp. 131–138. Association for computing machinery (2020). https://doi.org/10.1145/3414752.3414779
66. PWC: Fit to compete: Accelerating digital workforce transformation in financial services. https://www.pwc.com/gx/en/issues/upskilling/pwc-digital-upskilling.pdf (2019). Accessed 17 Apr 2024
67. Rathi, M.M., Parimalarani, G.: Impact of digital transformation on employment in banking sector. Int. J. Sci. Technol. Res. **9**, 4912–4916 (2020)
68. Redecker, C.: European framework for the digital competence of educators: DigCompEdu. In: Punie, Y. (ed.) EUR 28775 EN. Publication Office of the EU. https://doi.org/10.2760/159770 (2017). Accessed 4 June 2024
69. Shook, E.J., Knickrehm, M.A., Costonis, M.A., Woolf, A., Lavelle, K.: Insurers must reskill and reshape their workforces to seize growth opportunities from artificial intelligence, according to

research from Accenture. https://www.accenture.com/_acnmedia/pdf-79/accenture-insurance-report-ai-future-workforce-survey.pdf (2018). Accessed 10 Mar 2024
70. Shook, E.J., Knickrehm, M.A., Mcintyre, A., Woolf, A., Browne, Y., Lavelle, K.: Future workforce survey banking: Realizing the full value of AI. Accenture. https://www.readkong.com/page/full-value-of-ai-banking-realizing-3058273 (2018). Accessed 10 Mar 2024
71. Soukupová, N., Adamová, M., Krninská, R.: Industry 4.0: An employee perception (case of the Czech Republic). Ac. Univ. Agric. Et Sil. Mend. Brun. **68**, 637–644 (2020). https://doi.org/10.11118/actaun202068030637
72. UNESCO: UNESCO ICT competency framework for teachers. https://iite.unesco.org/pics/publications/en/files/3214694.pdf (2011). Accessed 12 Mar 2024
73. Velthuijsen, J.W., Van Tol, W., Hagen, A.: Human value in the digital age. PwC. https://www.pwc.nl/nl/assets/documents/pwc-human-value-in-the-digital-age.pdf (2017). Accessed 3 July 2024
74. Wellins, R.S., Sadjady, S.: HR under pressure: Falling behind in the race to transform. In: Sinar, E., Ray, R.L., Wellins, R.S., Neal, S., Canwell, A.L., Abel, A.L., Popiela, A., Dettmann, J., Rolland, L., Collins, L., Cotton, T. (eds.) The Global Leadership Forecast 2018. 25 Research Insights to Fuel Your People Strategy, pp. 50–51. Development Dimensions International, Conference Board, EYGM Limited, (2018)

Chapter 6
Empirical Part of Digitalization in the Context of HRM

Abstract Chapter 6 presents the main results of the primary research conducted as part of this book (Fig. 1.2 in Sect. 1.3). This Chapter 6 analyzes the primary data collected in the Czech Republic and abroad from 2019–2022 from the perspective of employers and employees operating in the selected sectors. It also aims to answer the research questions and test the hypotheses formulated in the theoretical part of the book (see Sect. 1.2).

6.1 Digitalization and Digital Competencies from the Employers' Perspective

Over the past few years, digital competencies have become a key topic in the debate about what kinds of knowledge and skills employees should have to work, learn, and engage effectively in modern society. The spread of DTs and information accelerated by the COVID-19 pandemic has significantly impacted the labor market and the competencies required by employers from their employees (see Sect. 5.1). Section 6.1 uses primary data to establish employers' views on what digital competencies employees need at work and the main barriers to learning and developing digital competencies.

6.1.1 Pilot Survey of Czech Enterprises

As mentioned in Sects. 1.5 and 1.6, pilot research on Czech companies on digitalization and its impact on employee competencies was conducted in September 2019 at FE TUL. It targeted SMEs representatives primarily from the ranks of top management and HR management. The pilot survey was aimed to investigate the interest of Czech enterprises in digitalization and the development of employees' digital competencies to gain a deeper understanding of it within the framework of this book and verify the appropriateness of the question setting for the respondents. Thirty

A. Mazurchenko, *The Impact of the Digitalization on the Human Resource Management*, Information Systems Engineering and Management 25,
https://doi.org/10.1007/978-3-031-76790-6_6

responses were obtained from randomly drawn respondents. It should also be noted that this data collection was conducted before the start of the COVID-19 pandemic. Subsequently, the course and the respondents' answers in subsequent surveys may already reflect the situation with the pandemic, which had a significant impact on HR digitalization.

In the pilot survey, a third of respondents represent SMEs, and more than half represent respondents from large enterprises with more than 250 employees.

In order to test the assumption that enterprise size by the number of employees does not affect the degree of digitalization, the Author of this book decided to include in the analysis both responses obtained from SMEs and large enterprises. In most cases, the questionnaire was completed by employees (i.e., 36.7%), with one in four respondents being a top management representative or department head, including the HRM department. Employees were from companies whose business focuses on wholesale and retail trade (16.7%), manufacturing (13.3%), and professional, scientific, and technical activities (13.3%). The main findings of this survey are presented below.

The results of the pilot survey show that the vast majority of the companies surveyed (96.7%) use DTs in their business daily, which confirms the relevance of the topic under study, in particular, IT and software, mobile and web applications (20.0%), as well as other unspecified modern technologies (20.0%). One in ten employers participating in the pilot survey perceive all DTs, including automation, robotics, social media, IoT, and AI, as necessary to their business.

The pilot survey found helpful information regarding the planned investment of businesses in DTs between 2020 and 2024. The survey was conducted in September 2019, when it was yet to be expected that the global pandemic COVID-19 would break out within a few months. It was confirmed that Czech employers surveyed planned to invest financial resources mainly in cyber security (46.7%), process and product innovation (43.3%), and cloud services (40.0%), including remote data management. Applications for employees' T&D in digital skills were also among employers' potential areas of great interest, with more than half of the respondents (56.7%) intending to invest in them. It is evidence that employers in the surveyed companies know the need to develop their HRs, which can vastly increase their business's competitiveness in the digital era.

Another key finding of the pilot survey was that DT is used most often in the surveyed companies for marketing and sales (70.0%), product and service development (63.4%), accounting and financial management (56.7%), or administration (53.4%). One in five respondents said that their company does not use DTs for HRM, which confirms that the potential of DTs in this area needs to be fully exploited.

It should also be pointed out that around three-quarters of the surveyed companies (73.3%) were engaged in systematic T&D of their employees in digitalization during the survey period on ICT skills (70.0%), problem-solving skills related to the use of DTs (50.0%) and the ability to learn and retrain quickly as a result of the progressive digitalization of the workplace (50.0%).

It should be noted that every second respondent in the pilot survey (i.e., 53.3%) does not consider the introduction of new DTs into the business to be a big challenge, given that their employees are already prepared to use DTs in their work and there is no need to develop them further. Half of the respondents believe that the automation and robotization of business processes should not reduce the number of their employees.

The Author of this book subsequently established and verified an assumption based on the pilot survey results. The objective was to determine whether there is a statistically significant relationship between respondents' reactions to the following statements:

- Variable 1: "I consider employees' problem-solving skills to be important due to the progressive digitalization and automation of the workplace";
- Variable 2: "We use DTs in production the most in our company";
- Variable 3: "In the next five years, my company intends to invest in process and product innovation".

Spearman's rank correlation coefficient was used to test this assumption under the following hypotheses:

H_0: The value of the Spearman's correlation coefficient is zero.

H_1: The Spearman's correlation coefficient is different from zero.

For Variables 1 and 2, the value of Spearman's rank correlation coefficient is statistically significantly different from zero at the 1% significance level and, therefore, at the 5% significance level (Table 6.1). It demonstrates a relatively weak direct dependence (i.e., the P-value of 0.009 is lower than both 0.01 and 0.05) and the well-understood tendency that employers who make more use of DTs in production within their business also tend to rate the problem-solving skills of employees as necessary due to the progressive digitalization and automation of the workplace.

As shown in Table 6.1, for Variables 2 and 3, Spearman's correlation coefficient is not statistically significantly different from zero at either 1% or 5% significance levels (i.e., the P-value of 0.548 is more significant than 0.01 and 0.05). There was no statistically significant (correlation) relationship between the two variables.

For Variables 1 and 3, Spearman's correlation coefficient is statistically significantly different from zero at the 5% significance level (a P-value of 0.047 is more significant than 0.01 and less than 0.05). Thus, a relatively weak direct dependence between the variables is found even though Spearman's correlation coefficient is, at

Table 6.1 Calculation of Spearman's correlation coefficient for three selected variables

Variables	Correlation	Variable 1	Variable 2	Variable 3
Variable 1	Spearman's correlation coefficient	1.000	0.469**	0.366*
	P-value	–	0.009	0.047
Variable 2	Spearman's correlation coefficient	0.469**	1.000	0.114
	P-value	0.009	–	0.548
Variable 3	Spearman's correlation coefficient	0.366*	0.114	1.000
	P-value	0.047	0.548	–

*The correlation is significant at the 0.01 level

**The correlation is significant at the 0.05 level

first sight, "quite distant" from the value of one. It demonstrates that respondents who intend to invest more in process and product innovation in the next five years are more likely to rate their employees' problem-solving skills as necessary due to the progressive digitalization and automation of the workplace.

The main limitation of the pilot survey is that it was targeted at a selected small group of representatives of Czech companies who are also thematically focused on digitalization (i.e., participants of the Knowledge Transfer Week at FE TUL). Given this, it is impossible to generalize the pilot survey results. It should be stressed that despite these limitations, the main objective of the pilot survey was fulfilled.

The pilot survey provided valuable insights and served as a primary information source for mapping the issues under study, adjusting the concept of further research, its tools, and respondent selection strategy. Testing the respondents' questions also verified their wording's suitability for further research activities. Based on the feedback from the respondents, the pilot questionnaire was subsequently modified and used for primary data collection to explore the topic of digitalization and the development of employees' digital competencies from the perspective of Czech employers. The collection that took place in subsequent phases among companies (quantitative and qualitative data) and HEIs already fell within the period of the COVID-19 pandemic.

6.1.2 Research in Construction, Automotive, Banking and Insurance Sectors

Due to the size of the database of business entities operating in the CR as of 1 June 2020, the in-depth research in digitalization was limited to selected sectors of the Czech economy based on the analysis of secondary data presented in Sect. 5.2.

The survey aimed to map the current situation in the HRM regarding digitalization and identify Czech companies' needs in terms of employees' digital competencies. The sectors included the automotive industry, banking and insurance, and construction.

The research included ten banks and insurance companies, 27 construction and 39 automotive companies. The detailed selection of respondents and data collection are elaborated in Sect. 1.5. The profiles of respondents, taking into account the size of enterprises by number of employees, their industry according to the CZ-NACE classification of economic activities, and the region in which they operate, are presented in Table 6.2.

Half of the construction industry respondents to this survey work in the marketing and sales departments (29.6%) and the administration departments (18.5%). In comparison respondents from the automotive industry are more often represented by the owners of these enterprises (20.5%) or work in the production and quality department (25.6%) or in the HR department (18.0%). The central regions where the surveyed enterprises operate are the capital city of Prague, where all surveyed

Table 6.2 Profile of respondents by industry, size and region of business location

	Automotive (%)	Construction (%)	Banking and insurance (%)
Regions of the CR			
The capital city of Prague	2.5	22.2	100.0
Central Bohemia region	15.4	7.4	–
South Bohemia region	5.2	–	–
Pilsen region	2.5	3.7	–
Karlovy Vary region	2.5	–	–
Ustí nad Labem region	7.8	7.4	–
Liberec region	33.4	11.1	–
Hradec Kralove region	12.9	3.7	–
Pardubice region	2.5	3.7	–
Vysočina region	2.5	11.1	–
South Moravian region	-	14.9	–
Olomouc region	2.5	11.1	–
Moravian-Silesian region	7.8	3.7	–
Zlín region	2.5	–	–
Total	100.0	100.0	100.0
Number of employees in enterprises			
1–9	15.4	33.3	–
10–49	30.8	48.2	–
50–249	17.9	7.4	30.0
250 or more	35.9	11.1	70.0
Total	100.0	100.0	100.0

banks and insurance companies belong, and the Liberec region (CR). Regarding the size of enterprises by number of employees, about half of the respondents in the construction and automotive industries (55.6% vs. 48.7%) are SMEs.

The results of the survey point to the fact that the use of DTs in everyday activities is becoming common in all companies participating in the Czech financial sector and in the majority of enterprises operating in the automotive and construction sectors (100.0% vs. 89.7% vs. 82.5%). A relatively small percentage of respondents, namely 14.8% from the construction and 10.3% from the automotive sector, confirmed that they only occasionally use DTs in the workplace, and no one mentioned that they do not use them at all in their company's practices.

The respondents' opinions on whether some jobs and tasks become redundant due to DTs use are strongly divided. Most respondents from banks and insurance companies agree with this statement (90.0%). About half of respondents in the automotive industry also agree with this statement (48.7%). For Czech construction companies, however, the results did not support the assumption that DTs would significantly contribute to the fact that some jobs and tasks are no longer needed in daily work (70.4%).

It is worth noting that employee competency development is more often part of the long-term HRM strategy in banks and insurance companies than in automotive and construction companies (90.0% vs. 66.6% vs. 37.0%). Approximately one-third of the respondents in the Czech construction industry (29.6%) do not see the need for further training of employees to use DTs effectively in their daily work.

Respondents in this survey were also asked about their companies' readiness to use DTs in the context of the COVID-19 pandemic. For this purpose, a scale from one to seven was used, where one indicates the digitalization of business processes, and seven indicates that the company is avoiding using DTs in its operations. All the banks and insurance companies interviewed confirmed that they are trying to respond to the barriers caused by the COVID-19 pandemic, with 30.0% of these financial institutions being fully or almost fully digitized, meaning that the impact of the COVID-19 pandemic on them has not been significant. However, a significant challenge facing the banking and insurance sector is the acceptance of DTs by employees and the associated change in organizational culture. Most construction and automotive companies have successfully implemented DTs in selected business processes and continue in this direction. An interesting finding is that one in four construction companies avoid using DTs, and none intend to introduce changes in the coming year. One in ten construction businesses already operate in cloud storage, collaborate effectively online, and have all their business processes set up digitally. As far as the automotive industry is concerned, the main barrier to DTs adoption is that DT is separate from the organizational culture, and the mindset of employees still needs to be adapted to it (12.8%).

Most banks and insurance companies that participated in this survey (90.0%) regularly use DTs to improve their employees' competencies related to task performance. One in two automotive companies (51.3%) does the same, while only 37.0% of construction companies do the same. Employer-provided training is most often used for this purpose in the financial and automotive sectors. In contrast, in the construction sector, this method of skills acquisition is also preferred by employees (70.0% vs. 69.2% vs. 60.0%).

Regarding the level of employees' digital competencies, these are estimated to be expected in more than half of the companies surveyed, regardless of sector. In addition, almost one in four employees in the automotive industry (23.0%) has a low level of digital competencies. In contrast, the proportion of such employees in the construction industry is significantly lower (14.8%). In the financial sector, respondents did not indicate that they had employees with low or no level of digital competence. Notably, only a tiny proportion of employees in all three sectors are characterized by advanced digital competencies.

The survey results show that representatives of all three sectors agree on which employees' digital competencies are a priority for the company's further successful development. The top three leaders are basic data entry and processing skills, data skills, and problem-solving skills arising from working in a digital environment.

Based on these empirical research results, RQ_1 and RQ_2 are answered from the employers' perspective (see Sect. 1.2).

Employers surveyed in the financial sector require complex data analysis, mathematical skills, and software development skills much more than in other industries. However, web development skills are optional for the employers surveyed in the construction sector, which is logical given the industry and its needs.

According to the employers involved in this survey, only 5.1% of employees in Czech automotive companies can boast advanced data skills. In comparison, their share reaches 14.8% in the construction sector and 20.0% in the financial sector. It is imperative, as data skills are crucial for the continued functioning of companies in all three sectors.

In all three sectors studied, barriers to developing digital competencies arise mainly due to the heavy workload of employees that answers RQ_4 from the perspective of academics at FE TUL and the selected German university (see Sect. 1.2). It is confirmed by 80.0% of the respondents from the financial sector, more than half of the respondents from the automotive industry (56.4%), and less than a third from the construction industry (29.6%). Also essential is the general reluctance of employees to learn new things and their internal resistance to change, which is experienced by 60.0% of respondents from banks and insurance companies, more than a third of automotive companies (38.5%), and a quarter of construction companies (25.9%). It is worth noting that another quarter of the construction companies involved in this research believe that employers who underestimate employee development should be considered. The banks surveyed and insurance companies do not see any problem with financing the T&D in digital competencies, and automotive companies did not experience any problems finding suitable external agencies to organize T&D activities.

The frequency of T&D in digital skills across the selected sectors indicates that one in three construction companies surveyed do not see the need to update employees' digital skills, with one in six automotive employers agreeing and one in ten employers in the financial sector. Meanwhile, half of the banks and insurance companies surveyed believe that completing annual digital skills training will be sufficient to ensure that employees perform effectively in a digital environment. While one in four automotive companies see a need for training activities to develop employees' digital skills every few years, one in five construction companies support continuous T&D. A third of the banks and insurance companies surveyed are also keen to provide continuous digital skills training for their employees.

The survey results show that in-house training is the most commonly used method to close the digital competency gap for existing employees in all three sectors surveyed. It is preferred by 90.0% of surveyed employers in the financial sector, 53.8% of automotive, and 40.7% of construction companies. It should be noted that 40.0% of banks and insurance companies and around a fifth of employers in the automotive and construction sectors (23.0% vs. 18.5%) plan to recruit new employees who already have the required digital competencies. There is a demand for external training in only one-third of the surveyed banks, insurance companies, and automotive companies. The employment of agency staff and outsourcing digital tasks are used occasionally in all sectors surveyed. Based on the above, RQ_3 is answered from the employers' perspective (see Sect. 1.2).

Employers in all three sectors were also asked to independently identify work areas where they felt the need to improve digital skills. The results showed that data entry, processing, and analysis (15.0%), data encryption, storage, and security (10.8%), online communication with employees (8.5%), and digital communication with customers (8.3%) are among the critical areas for further improvement. Nevertheless, 8.3% of respondents believe they can work with what they know and do not feel the need to improve their digital skills.

The findings of this survey confirm that HRM, digital competencies development, and related employee T&D in the surveyed Czech construction and automotive companies and financial institutions have been significantly affected by the COVID-19 pandemic. The digitalization process has been dramatically accelerated, and the transition of most employees to working from home has led to the need to improve their digital competence levels, which is an advantage for these employers. Based on the above findings, most employers successfully implement DTs in stand-alone business processes and do not plan to stop this process. The disadvantage is that the workload of employees and the need for work-life balance has increased due to the Covid-19 pandemic. At the same time, the internal resistance of employees to change often became an obstacle to accelerating the development of their digital competencies in the interviewed companies in the initial phase of using the new DTs in their daily work activities. Employers who interview their employees demand basic data skills, problem-solving skills in a digital environment, and effective online communication.

The low return rate (i.e., 2.0%) is the main limitation of this survey. It i s impossible to generalize the results to all economically active automotive and construction companies, banks, and insurance companies in the CR. Despite the efforts of the Author of this book to establish cooperation with other partners to gain support and spread awareness about this survey (see Sect. 1.5), the return rate of the questionnaires could not be significantly influenced. Another limitation is that the survey only focuses on the progress of digitalization in HRM in one time period in the automotive, construction, banking, and insurance sectors and needs to consider its status during the COVID-19 pandemic in other sectors of the Czech economy. Nevertheless, the results provide precious and timely insight into the current state of digitalization in Czech companies in the context of HR.

6.1.3 *Digitalization in the Manufacturing Industry During the COVID-19 Pandemic: A Case Study*

Three in-depth semi-structured interviews were conducted with HR experts in three limited liability companies (LLC) operating in the manufacturing industry from 19 October to 17 December 2020 in the Liberec region (CR) to support the results obtained during the primary data collection. The design of the in-depth interviews and the rationale for selecting respondents are presented in Sect. 1.5. The results of semi-structured interviews are presented in the case studies in [1]. The findings from these case studies show that all three companies, motivated by new working conditions during the COVID-19 pandemic, made significant progress toward digital HRM. It was also confirmed that the HR managers interviewed recognize the growing importance of digital competencies for employees and the need to incorporate them into organizations' competency models. It can be concluded that the role of HR managers in adapting employees to the changing work environment and corporate culture during the COVID-19 pandemic in the surveyed companies is irreplaceable.

6.1.4 Digitalization in Banking and Insurance: A Case Study of a Large Czech Bank

The data presented in Sect. 5.2 from the banking sector confirmed that it is a highly digitalized environment that develops the digital competencies of its employees. An in-depth semi-structured interview was conducted with three HRM experts in one of the largest Czech banks on 24 November 2020 to understand digitalization in the banking sector better. Based on the in-depth interview, a case study was created to show the effects of the COVID-19 pandemic on HRM in the financial sector and to confirm the results of the primary data collection presented in Sect. 6.1. The case study is published in [2].

Both case studies (see Sect. 6.1.3) showed that regardless of sector, to maximize the benefits of digitalization in the COVID-19 pandemic, it is vital to foster an organizational culture that embraces change. In HRM, the T&D of employees in digital competencies and data skills were shown to be key topics of focus in all companies interviewed. Based on the results of the empirical research presented in Sect. 6.1, RQ_5 is answered from the employers' perspective (see Sect. 1.2).

6.1.5 Digitalization and Digital Competency from the Employers' Perspective: Testing Statistical Hypotheses

Section 6.1.5 aims to test hypotheses H_1, H_2, and H_3 (see Sect. 1.2) based on primary data collected in the CR from 76 respondents. These were representatives of 39 automotive companies, 27 construction companies, and 10 banks and insurance companies (see Sect. 6.1). For this purpose, Pearson's chi-square independence test and ANOVA analysis were chosen (see Sect. 1.7). The results of testing hypotheses H_1, H_2, and H_3 are summarised in Table 6.3.

Table 6.3 Testing the statistical hypotheses H_1, H_2 and H_3 from the employers' point of view

Statistical hypothesis	Result of statistical testing
H_1: Employee competence level in terms of using DTs is independent of the sector in which the employer operates according to the CZ-NACE classification of economic activities	Hypothesis H_1 is not rejected/ has been confirmed
H_2: The frequency of improving employee competencies necessary for practical work in the digital environment is independent of the sector in which the employer operates, according to the CZ-NACE classification of economic activities	Hypothesis H_2 is not rejected/ has been confirmed
H_3: According to the CZ-NACE classification of economic activities, enterprises' readiness to use DTs in the COVID-19 pandemic is independent of the sector	Hypothesis H_3 is rejected/not confirmed

Hypothesis H_1

In order to test hypothesis H_1, the following assumptions were made:

H_0: Employee competence level in terms of using DTs is independent of the sector in which the employer operates according to the CZ-NACE classification of economic activities.

H_1: According to the CZ-NACE classification of economic activities, there is a relationship between the level of competence of employees in terms of using DTs and the sector in which the employer operates.

To test hypothesis H_1 using ANOVA, it was first necessary to test whether the assumption that selection comes from a normal distribution is met. The Shapiro–Wilk normality test in IBM SPSS 25 was used to test normality (Table 6.4).

Since the P-value of the Shapiro–Wilk test criterion is lower than the chosen significance level of 5% (Table 6.4), the data are significantly different from a normal distribution. Under these conditions, simple analysis of variance becomes unreliable.

Subsequently, it was also necessary to check whether the assumption of homoskedasticity was met. Levene's test for agreement of variances in IBM SPSS 25 was used to verify this (Table 6.5).

Considering that the P-value of Levene's test is 0.048, which is lower than the chosen significance level $\alpha = 0.05$, hypothesis H_0 of the agreement of variances is rejected.

Because the assumptions of normality and homoskedasticity were not met, Welch's test instead of ANOVA should be used to test the H_1 hypothesis (Table 6.6).

Table 6.4 Normality testing to test hypothesis H_1

Shapiro–Wilk test			
Level of employees' digital competencies	Value	Degrees of freedom	P-value
	0.698	74	0.000

Table 6.5 Testing homoskedasticity to test hypothesis H_1

Levene's test			
Level of employees' digital competencies	Value	Degrees of freedom	P-value
	3.161	2	0.048

Table 6.6 Verification of the statistical relationship between the level of employees' digital competencies and the economic sector in which the employer operates

Welch's test[a]			
Level of employees' digital competencies	Value	Degrees of freedom	P-value
	3.148	2	0.057

[a] Asymptotic F distribution

Since the P-value for Welch's test is higher than the chosen significance level of 5% (Table 6.6), there was no statistically significant relationship between the level of employees' competencies in terms of DTs use and the sector in which the employer operates according to the CZ-NACE classification of economic activities.

Hypothesis H_2

In order to test hypothesis H_2 using Pearson's chi-square test of independence, the following assumptions were made:

H_0: The frequency of improving employee competencies necessary for practical work in the digital environment is independent of the sector in which the employer operates, according to the CZ-NACE classification of economic activities.

H_1: There is a relationship between the frequency of improving employee competencies necessary for practical work in the digital environment and the sector in which the employer operates, according to the CZ-NACE classification of economic activities.

The test criterion value x^2 calculated in IBM SPSS 25 to test hypothesis H_2 equals 17.507 (Table 6.7). At the same time, the critical value x^2 for a significance level of $\alpha = 0.05$ for ten degrees of freedom is set at 19.31.

Given that for each of the twelve cells of the contingency table (66.7%), the expected (calculated) frequency is less than five, and the minimum expected frequency is 0.46, Pearson's x^2 test of independence is unreliable due to the small sample size. In this situation, it is better to use Fisher's exact test.

Table 6.7 Testing the statistical relationship between employees' level of competencies in terms of DTs use and industry

Statistical method	Value	Degrees of freedom	P-value
Pearson's chi-square test of independence	17.507[a]	10	0.064
Fisher's exact test	15.653	–	0.062

[a] Twelve cells (66.7%) have an expected number less than 5. The minimum expected count is 0.46.

Since the P-value for Fisher's exact test is higher than the chosen significance level of 5% (Table 6.7), no statistically significant relationship was shown between the frequency of improvement of employee competencies necessary for practical work in the digital environment and the sector in which the employer operates, according to the CZ-NACE classification of economic activities.

Hypothesis H_3

The following assumptions were made to test hypothesis H_3:

H_0: The level of readiness of enterprises to use DTs in the COVID-19 pandemic is independent of the sector according to the CZ-NACE classification of economic activities.

H_1: There is a relationship between enterprises' readiness to use DTs during the COVID-19 pandemic and the sector according to the CZ-NACE classification of economic activities.

To test hypothesis H_3 using ANOVA, it was first necessary to test whether the assumption that selection comes from a normal distribution is met. The Shapiro–Wilk normality test in IBM SPSS 25 was used to test normality (Table 6.8).

Since the P-value of the Shapiro–Wilk test criterion is lower than the chosen significance level of 5% (Table 6.11), the data are significantly different from a normal distribution. Under these conditions, simple analysis of variance becomes unreliable.

Subsequently, it is also necessary to verify whether the assumption of homoskedasticity is met. Levene's test for agreement of variances in IBM SPSS 25 was used to verify this (Table 6.9).

Given that the P-value of Levene's test is lower than the chosen significance level $\alpha = 0.05$ (Table 6.9), the hypothesis H_0 of agreement of variances is rejected.

Because the assumptions of normality and homoskedasticity were not met, it is appropriate to use Welch's test instead of ANOVA to test hypothesis H_3 (Table 6.10).

Since Welch's test's P-value is below the chosen significance level of 5% (Table 6.10), there is a statistically significant relationship between the level of

Table 6.8 Testing normality to test hypothesis H_3

Shapiro–Wilk test			
Business preparedness to use DTs during the COVID-19 pandemic	Value	Degrees of freedom	P-value
	0.804	76	0.000

Table 6.9 Testing homoskedasticity to test hypothesis H_3

Levene's test			
Business preparedness to use DTs during the COVID-19 pandemic	Value	Degrees of freedom	P-value
	8.786	2	0.000

Table 6.10 Verification of the statistical relationship between the level of preparedness of enterprises to use DTs in the COVID-19 pandemic and the economic sector

Welch's test[a]			
Business preparedness to use DTs during the COVID-19 pandemic	Value	Degrees of freedom	P-value
	3.809	2	0.030

[a]Asymptotic F distribution

Table 6.11 Verification of employer consensus in the construction and banking and insurance sectors on Statement 1

Order statistics			
Industries	Number of enterprises in each file	Average ranking of companies in each file	Sum of the rankings of the companies in each file
Construction	27	22.87	617.5
Banking and insurance	10	8.55	85.5
Total	37		
Summary statistics			
Replacing tasks and work with DTs			
Mann–Whitney U test	30.500		
P-value	0.000		

readiness of enterprises to use DTs during the COVID-19 pandemic and the sector according to the CZ-NACE classification of economic activities.

Mann–Whitney Test

The Author of this book has also established the assumption that the views of employers in the automotive, construction, banking, and insurance sectors regarding the following statement do not differ:

"The use of DTs may cause some tasks and jobs to be in your company" (Statement 1).

Statement 1 was tested using the Mann–Whitney U test, which is suitable for small samples and ordinal scales, and was based on the following hypotheses:

H_0: Views of employers in the automotive, construction, and building industries, as well as banking and insurance, do not differ.

H_1: Views of employers in the automotive, construction, and building industries and banking and insurance industries differ.

The P-value calculated in IBM SPSS Statistics 25 software is lower than the chosen significance level of the test $\alpha = 0.05$ in the statistical analysis of the responses obtained from construction banking and insurance sector employers (Table 6.11).

Similarly, the P-value calculated is lower than the chosen significance level of the test $\alpha = 0.05$ in the statistical analysis of the responses obtained (Table 6.12).

The same was confirmed in the case of the P-value calculation, where the calculated value was lower than the chosen significance level of the test $\alpha = 0.05$ in the statistical evaluation of the responses obtained from employers in the automotive and construction industries (Table 6.13).

Consequently, it can be concluded that the null hypothesis H_0 is rejected in favor of the alternative hypothesis H_1 in the case of all three sectors examined. Thus, a statistically significant difference was found in employers' attitudes in the automotive, construction, banking, and insurance industries toward the statement, "The use of DTs can make some tasks and jobs unnecessary in your company".

Table 6.12 Validating the consensus of employers' views in the automotive and banking and insurance industries on Statement 1

Order statistics			
Industries	Number of enterprises in each file	Average ranking of companies in each file	Sum of the rankings of the companies in each file
Automotive industry	39	27.82	1085.00
Banking and insurance	10	14.00	140.00
Total	49		
Summary statistics			
Replacing tasks and work with DTs			
Mann–Whitney U test	85.000		
P-value	0.004		

Table 6.13 Validating the consensus of views of employers in the automotive and construction industries on Statement 1

Order statistics			
Industries	Number of enterprises in each file	Average ranking of companies in each file	Sum of the rankings of the companies in each file
Automotive industry	39	29.41	1147.00
Construction	27	39.41	1064.00
Total	66		
Summary statistics			
Replacing tasks and work with DTs			
Mann–Whitney U test	367.000		
P-value	0.029		

6.2 Digitalization and Digital Competencies from the Employees' Perspective

As a result of the dynamic development of DTs, systematic mapping of the status of employees' digital competencies and their development is becoming more complex as the relevance of individual work activities in the digital environment and the associated competencies are constantly change. This became even more evident during the COVID-19 pandemic.

However, providing access to DTs does not automatically mean that employees will know how to use them correctly to enhance their work performance and meet the organization's objectives.

Section 6.2 aims to explore employees' perceptions of their digital readiness, define the digital competencies that employees believe are important in their work, and identify the main barriers they encounter in their learning and development in digital competencies.

6.2.1 Research on the Digital Readiness of Employees in Different Sectors

As detailed in Sects. 1.5 and 1.6, the digital readiness survey was conducted between April and June 2021. The respondents were employees from all years of CS at FE TUL working in companies from different sectors of the Czech economy.

The employee digital readiness survey explored employee attitudes toward digitalization and the development of digital competencies at work during the COVID-19 pandemic.

Thirty-one responses were obtained from randomly drawn respondents, and it can be concluded that the sample of employed FE TUL students of CS is representative. Descriptive statistics of the results obtained are presented below.

As can be seen from Table 6.14, this survey had the highest number of respondents aged between 21 and 30 (71.0%) who work as ordinary employees (68.3%) in large enterprises with more than 250 employees (48.4%) with headquarters in Liberec (38.8%) or Central Bohemia (22.5%) in the CR. In addition, most employees (93.6%) were employed full-time.

It is important to note that respondents were only willing to use DTs to support their education and development at FE TUL. On the contrary, most student employees (80.7%) are very interested. However, the importance of the different digital skills that studying at FE TUL contributes to the development of staff varies significantly - three-quarters of respondents reported the most outstanding development of their basic digital skills (e.g., working with MS Office word processing) and information skills (e.g., effective internet search, bibliographic citation of sources). On the contrary, only one in ten employees feel that FE TUL supports them in developing digital skills necessary for creating and operating a profile on professional social networks, safe

Table 6.14 Profile of respondents by age, job category, size, and region of employer location

Regions of the CR	Age of respondent (%)			Respondent's job position (%)			
	21–30 years old	31–40 years old	41–50 years old	An ordinary employee	Line management	Middle management	TOP management
Capital City of Prague	16.1	–	–	12.9	–	3.1	–
Central Bohemia Region	19.4	–	3.1	12.9	9.7	–	–
Ustí nad Labem Region	12.9	–	–	9.7	–	3.1	–
Liberec Region	12.9	19.4	6.5	26.3	3.1	6.5	3.1
Hradec Kralove Region	9.7	–	–	6.5	3.1	–	–
Total	1000			1000			
Number of employees in enterprises							
1–9	12.9	–	–	9.7	–	–	3.1
10–49	9.7	6.5	–	12.9	–	3.1	–
50–249	12.9	6.5	3.1	6.5	9.7	6.5	–
250 or more	35.4	6.5	6.5	38.9	6.5	3.1	–
Total	1000			1000			

online behavior, and managing their digital identity. These are skills that employees generally lack across sectors and professions in the COVID-19 pandemic.

The survey results also show that most employees rate their current level of skills in using DTs as advanced (58.1%) or intermediate (32.2%). Given the respondents' age (i.e., 20 to 50 years old), it is not surprising that all these employees have a low level of digital skills.

Overall, the employees surveyed use DTs very often (58.1%) or often (32.2%) at work. However, due to the COVID-19 pandemic, the intensity of individual digital tools and DTs at work has been affected. This was most evident in communication and collaboration, with one in two respondents (54.8%) confirming that they use DTs such as Zoom, Google Meet, MS Teams, Discord, and Slack much more frequently.

In contrast, virtually one in two respondents (48.4%) reported no changes in their use of digital time and appointment scheduling tools due to the coronavirus pandemic. Similarly, almost one-fifth of the employees surveyed (19.3%) have yet to start using digital tools for sharing and storing documents, either more or less. 45.1% of the employees in this survey do not use digital task and project management tools such as Trello, Orgpad, MS Planner, Asana, or Miro in their work at all.

Basic data skills (80.6%) and problem-solving skills in a digital environment (41.9%) are critical in their work. Software development skills are considered the most minor needed (3.2%). These results could be primarily influenced by the job position or department in which the employee works (see question 7). In terms of the learning and development support systems used by the employees surveyed, internal organizational learning systems predominate (67.7%). However, almost one in five employees (19.4%) use Google Classroom, and one in six (16.1%) use Moodle or no learning and development support system in the corporate environmentl. Based on the above research, RQ_1 is answered from the perspective of employees from different sectors (see Sect. 1.2).

Employer-provided training is the most commonly preferred way of acquiring the skills needed to use DTs at work (51.6%). A quarter of employees (25.8%) reported acquiring digital skills through informal help from colleagues and friends. Nevertheless, only a tiny proportion of employees (6.5%) take responsibility for their professional development and place the most significant emphasis on self-learning. Based on these empirical research results, RQ_3 is answered from the perspective of employees from different sectors (see Sect. 1.2).

Nine out of ten employees (90.4%) in this survey agreed that their employer should invest more in developing their digital skills. When asked what hinders employees in developing their digital competencies at work, the answer was the lack of time due to workload (38.7%), which answers RQ_4 from the perspective of employees from different sectors (see Sect. 1.2).

Lack of technical support at the workplace is also a significant barrier (32.3%), and the content of a digital training course that an employee would be willing to take must match the employer's needs (32.3%). Even though a lack of motivation to take digital training is not a challenge, one in six employees surveyed (16.1%) do not realize why they should take it.

Only a quarter of employees surveyed (25.8%) know the need for ongoing T&D in digital skills. More than a third of respondents in this survey (38.7%) update their digital skills at least once every six months to ensure adequate performance in a digital work environment. 16.1% of respondents do not see the need to update their digital skills at all.

A critical limitation of the survey is that it does not consider the views of employees under 20 and over 50 on digitalization and the development of their digital competencies and covers a limited geographical area (i.e., only 5 out of 14 regions of the CR).

Another limitation is that more information needs to be provided on the respondents' job titles, the department in which they work, or the sector in which their employer operates according to the CZ-NACE classification of economic activities. It could affect the interpretation of the results, given that different jobs require different levels of digital readiness of employees.

6.2.2 *Research on Digital Readiness of Employees of Selected Universities in the Czech Republic and Germany*

Given the significant impact of the COVID-19 pandemic and the increasing demands for the use of digital tools in teaching and collaboration with colleagues in academic settings (see section D in Sect. 5.2), this section aims to present the main findings on the impact of digitalization on the work of and competency profiles of employees in the environment of a selected one Czech and one foreign HEI and to compare the results obtained with each other. Table 6.15 summarises the essential characteristics of the surveyed HEIs regarding the digital readiness of their staff.

This survey aimed to assess the readiness of academic staff at FE TUL. It selected HEI in Germany during the COVID-19 pandemic and identify the main challenges these academics face in their work. Primary data collection occurred at FE TUL in the summer of 2021, when 100 employees (i.e., 85 academic staff and 15 non-academic staff) were working at the faculty. The return rate of the questionnaires among academics was 46.25% (i.e., 37 responses).

Primary data collection at the selected German university occurred in the spring of 2022 (see Sect. 1.6) when 153 academics worked there. The subsequent return rate was 55.6%.

The German university's name has been anonymized at the request of its rector. As all questions in the questionnaire were optional, Table 6.16 shows the responses obtained for each question at the German university.

Table 6.17 presents the profile of respondents, taking into account their age and length of employment at the faculty or university.

The survey results at FE TUL show that virtually all employees consider the development of skills related to digitalization to be significant (94.6%), which is in line with the strategic plan of the faculty for the period 2021–2025. During the research

Table 6.15 Comparison of the characteristics of the HEIs surveyed

Classification criteria	FE TUL	German university
Data collection period	24 June - 22 July 2021	28 April - 15 May 2022
Number of respondents	80 academic staff	153 academic staff
The most represented age group	More than 50 years (46.0%)	40–50 years (41.8%)
Digital tools and DTs used in teaching	Google Meet, E-learning, Google Forms, Google Jamboard, MS Teams, Zoom, Kahoot, Padlet, Slido, Google Classroom, Mentimeter	BigBlueButton, OPAL, Filr, ONYX, Nextcloud, Videocampus, Matrix, Survey, Jitsi, AdobeConnect, DFN Conf
Digital tools and DTs for communication	Google Meet, Zoom, MS Teams, Discord, Slack	BigBlueButton, Zoom, MS Teams
Digital tools and DTs for document sharing	MS OneDrive, Google Drive, DropBox	Nextcloud, DropBox, MS OneDrive
Digital tools and DTs for task and project management	73.0% of respondents do not use at all	Trello, Miro, OPAL
Digital tools and DTs for scheduling meetings and time	Google Calendar, One Note, Timely	Group Wise, MS Outlook, Doodle, DFN-Terminplaner
Most in-demand digital skills	Digital skills needed for teaching (86.5%), basic data skills (54.1%), complex data analysis and mathematical skills (37.8%), digital project management skills (32.4%)	Digital skills needed for teaching and communicating with students (88.9%), digital content creation skills (85.2%), using DTs for methodical design (56.8%), online protection skills (40.7%)
Level of digital skills	Intermediate (54.1%) and advanced (37.8%)	Advanced (50.6%) and competent (29.4%)
The most common ways of acquiring digital competencies	Self-study, informal help from colleagues and friends, in house training provided by the employer	Self-study, offers from the university's Centre for E-Learning and from faculty, informal help from colleagues and friends
Main obstacles in the development of digital competencies	Staff workload, lack of technical support in the workplace and motivation to undergo training in digitalization	Staff workload, other priorities, lack of targets for further training in digitalization

Table 6.16 Frequency of answers to each question at the German university of applied sciences

	Question number in the questionnaire															
	1	2	3	4	5	6	7	8	9	10	11	12	13	14	15	16
Number of responses from respondents	85	85	85	81	81	81	61	59	81	81	46	81	79	79	79	79

Table 6.17 Profile of respondents by age and length of employment

Respondents	Age of respondent (%)					Length of employment (%)				
	31–40 years old	41–50 years old	41–50 years old	51–60 years old	No answers	Less than 1 year	1–5 years	6–10 years	11–20 years	More than 20 years
FE TUL	24.3	27.0	27.0	–	2.7	2.7	5.4	10.8	40.5	40.5
German University	7.6	41.8	29.1	7.6	13.9	3.8	15.2	27.8	32.9	20.3

period at the German university, the strategic plan for the university's development for the next period, given the impact of the COVID-19 pandemic, was discussed. Regardless, most academic staff at the German university (87.6%) perceived the development of their digital competencies as important in advancing digitalization.

It should be noted that the two questionnaires used different scales to measure the level of employees' digital competencies, which does not allow a comparison of the results at the Czech and German universities. However, it is essential to note that half of the academic staff at FE TUL rate their level of digital competence as intermediate (54.1%), and more than a third (37.8%) consider it advanced. In contrast, at the German HEI, half of the respondents noted that their digital competencies were advanced (37.8%), and less than a third considered their level of digital competencies competent (29.4%). It was found that 80.0% of the interviewed academic staff at the German university and 97.3% of the academic staff at FE TUL have the potential to develop their digital skills in the future at an expert level, considered the highest at both universities.

At Czech and German HEIs, most academic staff (81.1% vs. 85.9%) use DTs because their work requires it. It was required during the research due to the pandemic and the move to online teaching. Nevertheless, almost two-thirds of respondents (70.3% vs. 64.7%) are allowed by their employers to use what DTs they use in their academic activities. At both Czech and German universities (67.5% vs. 75.3%), most academic staff appreciate the benefits that the introduction of DTs brings to their work. Moreover, every second academic interviewed in both university environments (59.4% vs. 52.4%) likes to share valuable tips with colleagues about using DTs at work. However, this exchange was limited during the COVID-19 pandemic. It can be noted that not all respondents in both countries who use DTs (81.1% vs. 85.9%) see it as beneficial to their work (only 67.5% vs. 75.3%) or feel enough freedom in their choice of these technologies (70.3% vs. 64.7%).

Interesting findings occur when assessing the change in intensity in DT use at work due to the COVID-19 pandemic. Unsurprisingly, academics at both educational institutions started us ing DTs for communication and collaboration more frequently during the pandemic (93.8% in the CR vs. 94.6% in Germany). At a point in time when 67.9% of the interviewed German academic staff reported an increase in the use of DTs for task management and project management, it became clear that practically three-quarters of the academic staff at FE TUL (73.0%) did not use Trello, Orgpad, Aasana, Miro or MS Planner at all even in the period before the coronavirus pandemic. Due to the pandemic, academic staff at FE TUL started to use DTs for sharing and storing documents much more than their German colleagues (Fig. 6.1).

In addition, respondents from FE TUL report that the pandemic has forced them to significantly increase the frequency of using digital tools and DTs to plan meetings and work time, which every German respondent interviewed confirmed (Fig. 6.2).

The comparison of the tools used in teaching during the first year of the COVID-19 pandemic found that entirely different digital tools were chosen for this purpose and applications at the two HEIs. As can be seen in Fig. 6.3, at FE TUL, Google Meet was the most preferred (91.9%), the university-wide internal e-learning system for sharing learning materials with students (86.5%), and Google Forms (48.6%).

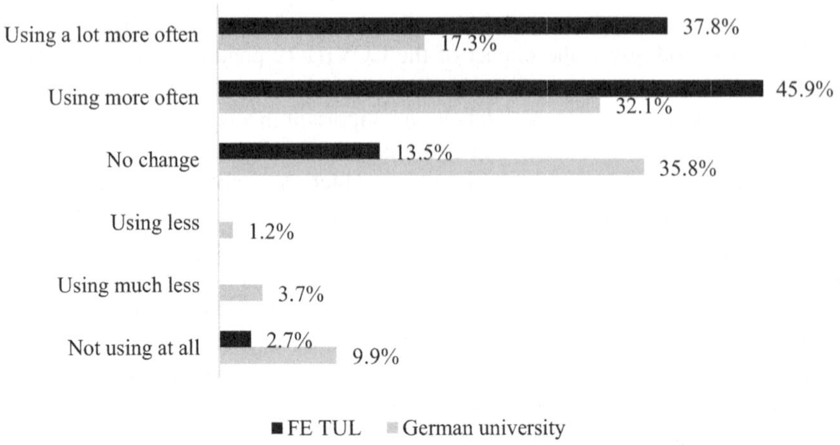

Fig. 6.1 Change in intensity of use of DTs for document sharing and storage during the COVID-19 pandemic

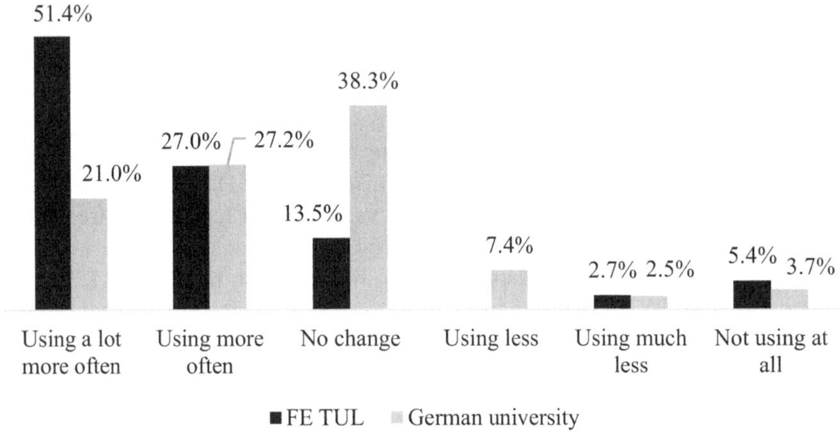

Fig. 6.2 Change in intensity in the use of DTs for appointment and time scheduling during the COVID-19 pandemic

Due to stricter requirements for student privacy, BigBlueButton (96.3%), OPAL (86.4%), and FLIR (42.0%) were the most commonly used for teaching at German HEI, as shown in Fig. 6.4.

Respondents from both HEIs disagreed on what digital skills are most needed for academic work.

As Fig. 6.5 shows, only one in five German respondents know they need skills to identify and solve technical problems with digital resources or DTs and to manage their digital profile and online identity.

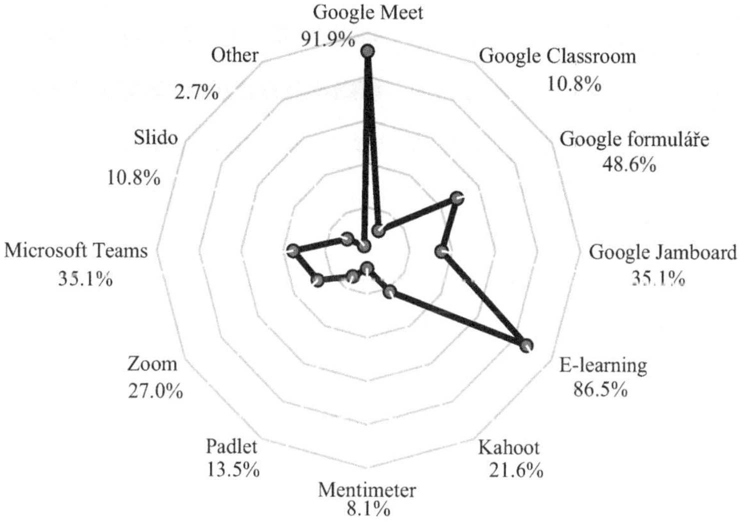

Fig. 6.3 Digital tools used for teaching in the pandemic period at FE TUL

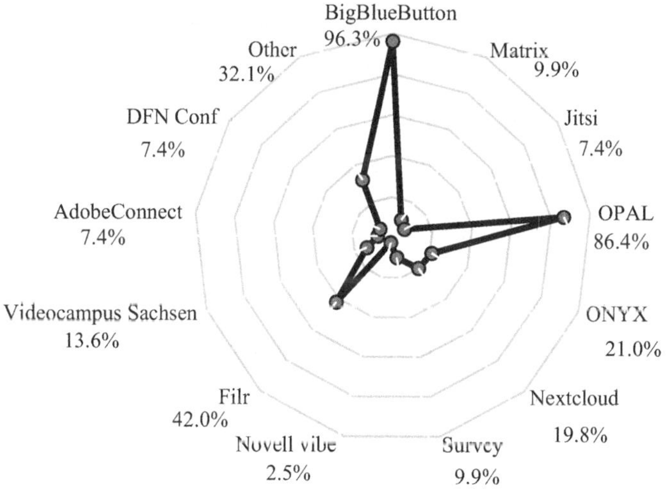

Fig. 6.4 Digital tools used for teaching in the pandemic period at the German university

It suggests that academic staff need regular technical support in these areas in the workplace. In addition, using DTs to communicate with other academic staff is rare, possibly due to a preference for face-to-face communication.

Digital teaching skills are also becoming a priority for academics at FE TUL, following the need to teach online during a pandemic (Fig. 6.6).

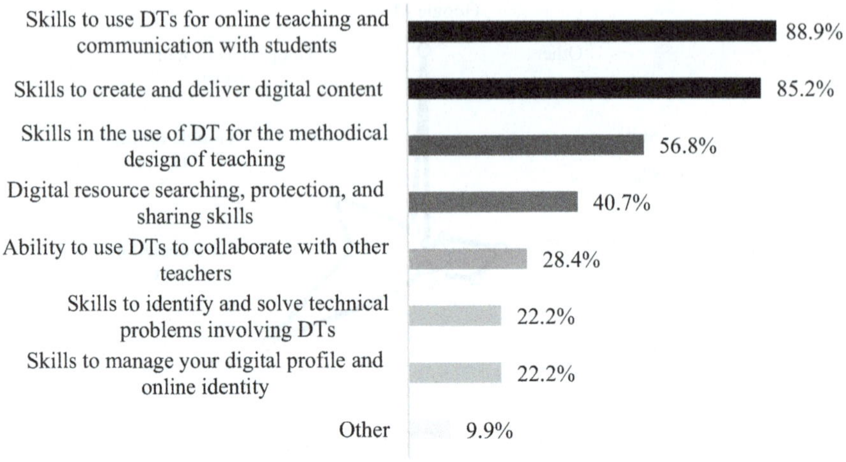

Fig. 6.5 Most needed digital skills, according to German academics

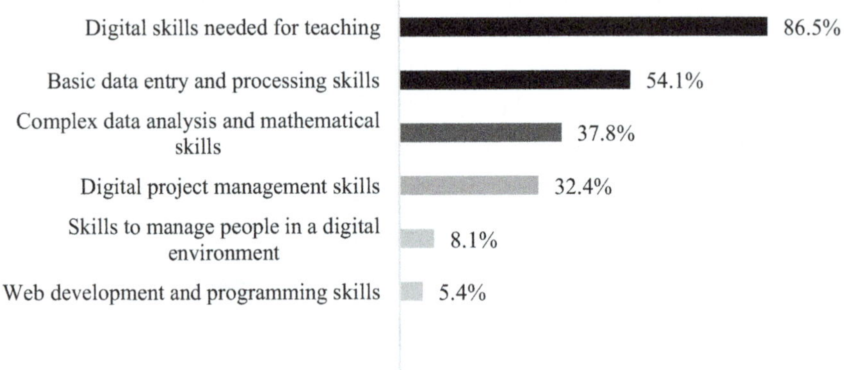

Fig. 6.6 Most needed digital skills, according to Czech academics

In addition, every second respondent stresses the importance of working with data, which answers RQ_1 from the perspective of academics at FE TUL and the selected German university (see Sect. 1.2).

In terms of the methods of acquiring digital skills by academics at a German university, self-study (55.6%), the offerings of the university's Centre for eLearning (33.3%), and the educational offerings that appear at individual faculties (29.6%) are frequently used. In contrast, in most cases, external training paid for by academics (79.0%) or by the employer (66.7%) is not used at all.

Regarding the frequency of digital competency training , half of the academic staff at FE TUL completes their training continuously or at least once a week (51.4%)

and almost one-fifth (18.9%) at least once every three months. Only a few respondents (8.1%) noted that they do not need to refresh digital competencies to perform effectively in the workplace.

At the same time, academics at FE TUL often resort to self-study (79.3%) when they need to develop their digital skills during a pandemic period. Like their German colleagues, they do not use external training when they have to pay for it themselves (89.2%) or even when funded by their employer (75.7%). Based on these empirical research results, RQ_3 is answered from the perspective of academics at FE TUL and the selected German university (see Sect. 1.2).

Two-thirds of the German academics surveyed (67.9%) need more time to learn and develop digital competencies due to their workload, and half (35.8%) prioritize other priorities. Similarly, 75.7% of academic staff at FE TUL refer to this significant barrier.

At the same time, one-third of the academic staff at FE TUL feel a need for more faculty support and technical support at the workplace (35.1% each), which is not a significant barrier for academics in Germany. The lack of motivation and goals for further training in digitalization is encountered by practically the same percentage of employees at both universities surveyed (16.2% vs. 18.5%). It can be concluded that academic staff at the selected universities in both the CR and Germany are fully aware of the need for further development and training in digital competencies during the pandemic period, as only a few of them stated that they do not need training in digital competencies (5.4% vs. 2.5%). Based on the above, RQ_4 is answered from the perspective of academic staff at FE TUL and the selected German university (see Sect. 1.2).

Based on the above analysis, the academic staff at FE TUL and the selected German university face similar challenges in developing their digital competencies. It is also worth noting that most respondents know the importance of digital competencies during the COVID-19 pandemic and the need for continuous development.

6.2.3 Digitalization and Digital Competencies from the Employees' Perspective: Testing Statistical Hypotheses

This section aims to test hypotheses H_4, H_5, and H_6 (see Sect. 1.2) based on primary data collected from representatives of 31 employees in different sectors (see Sect. 7.2), 37 academics at FE TUL, and 85 academics at a selected German university (see Sect. 7.2). For this purpose, Pearson's chi-square test of independence and ANOVA were chosen (see Sect. 1.7).

The results of testing hypotheses H_4, H_5, and H_6 are summarised in Table 6.18.

Hypothesis H_4

In order to test hypothesis H_4, the following assumptions were made:

Table 6.18 Testing statistical hypotheses H_4, H_5 and H_6 from the employees' point of view

Statistical hypothesis	Result of statistical testing
H_4: Employees' level of competence in using DTs at work is independent of their age	Hypothesis H_4 is not rejected/ has been confirmed
H_5: The frequency of improving employees' competencies in using DTs at work is independent of their age	Hypothesis H_5 is not rejected/ has been confirmed
H_6: Employee competence level in using DTs is independent of their length of employment in the company	Hypothesis H_6 is not rejected/ has been confirmed

H_0: Employees' competence level in using DTs at work is independent of their age.

H_1: There is a relationship between employees' levels of competence in using DTs at work and their age.

Employees from Different Sectors

First, to test Hypothesis H_4 using ANOVA, it was necessary to check whether the assumption that selection comes from a normal distribution is met$_4$. The Shapiro–Wilk normality test in IBM SPSS 25 was used to test for normality (Table 6.19).

Since the P-value of the Shapiro–Wilk test criterion is lower than the chosen significance level of 5% (Table 6.19), the data are significantly different from a normal distribution. Under these conditions, simple analysis of variance becomes unreliable.

Subsequently, it was also necessary to check whether the assumption of homoskedasticity was met. Levene's test for agreement of variances in IBM SPSS 25 was used to verify this (Table 6.20).

Given that the P-value of Levene's test is higher than the chosen significance level $\alpha = 0{,}05$, H_0 on the agreement of variances is not rejected (Table 6.20).

Table 6.19 Normality testing to test hypothesis H_4 on a sample of employees from different industries

Shapiro–Wilk test			
Level of employees' digital competencies	Value	Degrees of freedom	P-value
	0.783	31	0.000

Table 6.20 Testing homoskedasticity to test hypothesis H_4 on a sample of employees from different industries

Levene's test			
Level of employees'digital competencies	Value	Degrees of freedom	P-value
	0.519	2	0.601

As a result of the normality assumption not being met, it is appropriate to use the Kruskal–Wallis H test instead of ANOVA to test the H_4 hypothesis (Table 6.21).

In the context that the P-value of the Kruskal–Wallis H test is higher than the chosen significance level $\alpha = 0{,}05$, H_0 is accepted (Table 6.21). Consequently, there was no statistically significant relationship between the level of competence of employees working in different sectors regarding DT use and their age.

FE TUL Staff

First, it was necessary to test whether the assumption that selection comes from a normal distribution is met to test hypothesis H_4 using ANOVA. The Shapiro–Wilk normality test in IBM SPSS 25 was used to test normality (Table 6.22).

Since the P-value of the Shapiro–Wilk test criterion is lower than the chosen significance level of 5% (Table 6.22), the data are significantly different from a normal distribution. Under these conditions, simple analysis of variance becomes unreliable. Subsequently, it is also necessary to verify whether the assumption of homoskedasticity is met. Levene's test for agreement of variances in IBM SPSS 25 was used to verify this (Table 6.23).

Given that the P-value of Levene's test is higher than the chosen significance level $\alpha = 0{,}05$, H_0 on the agreement of variances is not rejected (Table 6.23).

As a result of the assumption of normality not being met, the Kruskal–Wallis H test is appropriate (Table 6.24).

Since the P-value of the Kruskal–Wallis H test is higher than the chosen significance level level $\alpha = 0{,}05$ (see Table 6.24), it can be concluded that hypothesis H_0 is accepted. Consequently, there was no statistically significant relationship between the level of competence of FE TUL employees in terms of DT use and their age.

German university staff

To test hypothesis H_4 using ANOVA, it was first necessary to test whether the assumption that selection comes from a normal distribution is met. The Shapiro–Wilk normality test in IBM SPSS 25 was used to test normality (Table 6.25).

Table 6.21 Verification of the statistical relationship between the level of digital competencies of employees from different sectors and their age

Kruskal–Wallis H test			
Level of employees' digital competencies	Value	Degrees of freedom	P-value
	1.780	2	0.411

Table 6.22 Normality testing to test hypothesis H_4 on a sample of FE TUL employees

Shapiro–Wilk test			
Level of employees' digital competencies	Value	Degrees of freedom	P-value
	0.767	36	0.000

Table 6.23 Testing homoskedasticity to test hypothesis H_4 on a sample of FE TUL employees

Levene's test			
Level of employees' digital competencies	Value	Degrees of freedom	P-value
	0.421	2	0.660

Table 6.24 Verification of the statistical relationship between the level of digital competencies of FE TUL employees and their age

Kruskal–Wallis H test			
Level of employees' digital competencies	Value	Degrees of freedom	P-value
	2.532	2	0.282

Table 6.25 Normality testing to test hypothesis H_4 on a sample of employees at a selected German university

Shapiro–Wilk test			
Level of employees' digital competencies	Value	Degrees of freedom	P-value
	0.825	68	0.000

Since the P-value of the Shapiro–Wilk test criterion is lower than the chosen significance level of 5% (Table 6.25), the data are significantly different from a normal distribution. Under these conditions, simple analysis of variance becomes unreliable.

Subsequently, it is also necessary to verify whether the assumption of homoskedasticity is met. Levene's test for agreement of variances in IBM SPSS 25 was used to verify this (Table 6.26).

Given that the P-value of Levene's test is higher than the chosen significance level $\alpha = 0.05$, H_0 on the agreement of variances is not rejected (Table 6.30).

Because the assumption of normality was not met, the H_4 hypothesis should be tested using the Kruskal–Wallis H test instead of ANOVA (Table 6.27).

Since the P-value of the Kruskal–Wallis H test is higher than the chosen significance level $\alpha = 0.05$ (Table 6.27), it can be concluded that hypothesis H_0 is accepted. Consequently, no statistically significant relationship was shown between employees' competence level at the selected German university in terms of the use of DTs and their age.

Table 6.26 Testing homoskedasticity to test hypothesis H_4 on a sample of employees at a selected German university

Levene's test			
Level of employees' digital competencies	Value	Degrees of freedom	P-value
	0.291	3	0.832

Table 6.27 Verification of the statistical relationship between the level of digital competencies of employees of a selected German university and their age

Kruskal–Wallis H test			
Level of employees' digital competencies	Value	Degrees of freedom	P-value
	3.048	3	0.384

Hypothesis H_5

The following assumptions were made to test this hypothesis using Pearson's chi-square test of independence:

H_0: The frequency of improving employees' competencies using DTs at work depends on age.

H_1: There is a relationship between the frequency of improving employees' competencies using DTs at work and their age.

Employees from Different Sectors

First, the value of the test criterion was calculated in IBM SPSS 25 (Table 6.28). Subsequently, the critical value x^2 was determined for a significance level of $\alpha = 0.05$, knowing the number of degrees of freedom. This critical value is equal to 19.31. Since the P-value equals 0.250, higher than the selected % significance level of 5%, the null hypothesis H_0 is not rejected.

Since for each of the 17 cells of the contingency table (94.4%), the expected frequency is less than five, and the minimum expected frequency is 0.19, Pearson's x^2 test of independence is unreliable. In this situation, it is better to use Fisher's exact test. Since the P-value for Fisher's exact test is higher than the chosen significance level of 5% (Table 6.28), there is no statistically significant relationship between the frequency of improvement in the competence of employees from different sectors in terms of the use of DTs at work and their age.

FE TUL Staff

In order to test hypothesis H_5, the value of the test criterion was first calculated x^2 (Table 6.29). Subsequently, the critical value x^2 was determined for a significance level of $\alpha = 0.05$, knowing the number of degrees of freedom. This critical value is equal to 15.51. Since the P-value is equal to 0.785, higher than the selected % significance level of 5%, the null hypothesis H_0 is not rejected.

Since for each of the ten cells of the contingency table (83.3%), the expected (calculated) frequency is less than 5, and the minimum expected frequency is 0.10,

Table 6.28 Verification of the statistical relationship between the frequency of improvement of competencies in terms of the use of DTs by employees from different sectors and their age

Statistical method	Value	Degrees of freedom	P-value
Pearson's chi-square test of independence	12,541[a]	10	0,250
Fisher's exact test	11,714	–	0,146

Table 6.29 Verification of the statistical relationship between the frequency of improvement of competencies in terms of the use of DTs by FE TUL employees and their age

Statistical method	Value	Degrees of freedom	P-value
Pearson's chi-square test of independence	3,471[a]	6	0,748
Fisher's exact test	5,041	–	0,666

Table 6.30 Normality testing to test hypothesis H_5 on a sample of FE TUL employees

Shapiro–Wilk test			
Level of employees' digital competencies	Value	Degrees of freedom	P-value
	0.789	37	0.000

Pearson's x^2 test of independence is unreliable due to the small sample size. In this situation, it is better to use Fisher's exact test. Since the P-value for Fisher's exact test is higher than the chosen significance level of 5% (Table 6.29), no statistically significant relationship was shown between the frequency of improvement in FE TUL employees' competence in terms of using DTs at work and their age.

The above results do not suggest a statistically significant relationship between the frequency of improving of employees' digital competencies and their age. Hypothesis H_5 could not be tested based on the responses of the German academics because the questionnaire did not ask about the frequency of improving their digital competence at work.

Hypothesis H_6

Hypothesis H_6 was also tested on a group of academic staff at FE TUL and a selected German university. To this end, the following assumptions were made:

H_0: FE TUL employees' competence in using DTs at work is independent of the length of their employment at FE TUL.

H_1: There is a relationship between the level of competence of FE TUL employees in using DTs at work and the length of their employment at FE TUL.

FE TUL Staff

To test hypothesis H_6 using ANOVA, it was first necessary to test whether the assumption that selection comes from a normal distribution is met. The Shapiro–Wilk normality test in IBM SPSS 25 was used to test normality (Table 6.30).

Since the P-value of the Shapiro–Wilk test criterion is lower than the chosen significance level of 5% (Table 6.30), the data are significantly different from a normal distribution. Under these conditions, simple analysis of variance becomes unreliable.

Subsequently, it is also necessary to verify whether the assumption of homoskedasticity is met. Levene's test for agreement of variances in IBM SPSS 25 was used to verify this (Table 6.31).

Table 6.31 Testing homoskedasticity to test hypothesis H_6 on a sample of FE TUL employees

Levene's test			
Level of employees' digital competencies	Value	Degrees of freedom	P-value
	3.206	3	0.036

Given that the P-value of Levene's test is lower than the chosen significance level $\alpha = 0{,}05$ (Table 6.31), the hypothesis H_0 of the agreement of variances is rejected.

Because the assumptions of normality and homoskedasticity were not met, Welch's test instead of ANOVA was appropriate to test the H_6 hypothesis (Table 6.32).

Since Welch's test's P-value is higher than the chosen significance level of 5% (Table 6.32), no statistically significant relationship was shown between the frequency of improvement in the competence of FE TUL employees in terms of using DTs at work and their age.

German University Staff

In order to test hypothesis H_6 using ANOVA, it was first necessary to verify that the assumption that the selection comes from a normal distribution is met. The Shapiro–Wilk normality test in IBM SPSS 25 was used to test normality (Table 6.33).

Since the P-value of the Shapiro–Wilk test criterion is lower than the chosen significance level of 5% (Table 6.33), the data are significantly different from a normal distribution. Under these conditions, simple analysis of variance becomes unreliable. Consequently, it is also necessary to check whether the assumption of homoskedasticity is met. The Levene's test for the goodness of fit of the variances in IBM SPSS 25 was used for verification (see Table 6.34).

Given that the P-value of Levene's test is higher than the chosen significance level $\alpha = 0{,}05$ (Table 6.34), H_0 on the agreement of variances is not rejected.

Table 6.32 Verification of the statistical dependence between the level of digital competencies of FE TUL employees and the length of their employment at FE TUL

Welch's test[a]			
Level of employees' digital competencies	Value	Degrees of freedom	P-value
	0.577	3	0.662

[a] Asymptotic F distribution

Table 6.33 Normality testing to test hypothesis H_6 on a sample of employees at a selected German university

Shapiro–Wilk test[a]			
Level of employees' digital competencies	Value	Degrees of freedom	P-value
	0.826	79	0.000

[a] Lilliefors test

Table 6.34 Testing homoskedasticity to test hypothesis H_6 on a sample of employees at a selected German university

Levene's test			
Level of employees' digital competencies	Value	Degrees of freedom	P-value
	1.781	4	0.142

Table 6.35 Verification of the statistical relationship between the level of digital competencies of the employees of a selected German university and their length of employment at this university

Kruskal–Wallis H test			
Level of employees' digital competencies	Value	Degrees of freedom	P-value
	1.465	4	0.833

Because the assumption of normality was not met, it is appropriate to use the Kruskal–Wallis H test instead of ANOVA to test the H_6 hypothesis (Table 6.35).

Since the P-value of the Kruskal–Wallis H test is higher than the chosen significance level $\alpha = 0{,}05$ (Table 6.35), the hypothesis H_0 is not rejected. Consequently, no statistically significant dependence was shown between the frequency of improvement in the competence of German university employees in terms of DT use at work and their age.

The above results do not suggest a statistically significant relationship between the frequency with which academic staff improve their digital competence and the length of their employment at the faculty or university.

6.3 Digitalization and Digital Skills: A Comparison of Employers' and Employees' Views

During the empirical research, the Author of this book compared the opinions of employers and employees on digitalization and digital skills in selected samples of respondents. Employee respondents were asked to identify the areas of their work in which they felt the most remarkable lack of digital skills during the COVID-19 pandemic. Similarly, employers were asked to specify three topics or areas requiring employees to improve their digital skills. On the employers' side, 39 representatives of companies from the automotive industry, 27 from construction companies, and 10 representatives of banks and insurance companies in the CR (i.e., 76 respondents) gave their opinions. On the employees' side, 31 randomly selected respondents working in various sectors in the CR expressed their opinions on the topic under study. All open answers were interpreted and grouped into individual areas of interest expressed by employers and employees in terms of digital skills development in which they were most interested (Table 6.36).

Table 6.36 Areas of digital skills development at work by Czech employees and employers

Areas for developing digital skills	Employers (%)	Staff (%)
Data entry, processing and analysis	15.0	12.7
Data encryption, storage and security	10.8	2.8
Digital communication with colleagues and customers	8.3	7.0
Creating and sharing digital content	6.7	5.6
Solving problems related to the use of DTs	6.7	9.9
Work with text editors and spreadsheet programs	5.0	7.0
Programming, website development and administration	5.0	9.9
Online marketing and advertising	2.5	4.2
Scheduling time and tasks	–	4.2
Work in a cloud environment	3.3	–
Social network management, digital identity	–	4.2
Basic graphic skills/graphics	–	2.8
Work in SAP	–	2.8
Other areas	28.4	18.4
No areas	8.3	8.5
Total	100.0	100.0

Based on the results presented in Table 6.36, data handling is considered a priority among respondents in terms of data processing and secure storage and sharing. One in four employers surveyed agree (i.e., 25.8%), and every sixth employee (i.e., 15.5%). The same emphasis is placed on the need for online communication and sharing of digital materials in a pandemic period by employers and employees (15.0% vs. 12.6%). It should be highlighted that employees perceive the need for programming and web management skills and the ability to solve problems in a digital environment more than employers require them to (9.9% vs. 5.0% and 9.9% vs. 6.9%).

It is also important to highlight that around a quarter of employers (28.4%) and a fifth of employees (18.4%) involved in this empirical research identified other areas for digital skills development that, due to their diversity, could not be classified into individual groups. It may be due to the specific nature of the sector in which the employers operate or the job positions of the employees.

Advantages and disadvantages in the work of academics in the context of digitalization during the COVID-19 pandemic

The book's Author also investigated the advantages and disadvantages of digitalization in academic work during the COVID-19 pandemic. For this purpose, academics from FE TUL and selected German universities were contacted, which allowed the interpretation of the results of international comparison. As all questions in the questionnaire at the selected German university were optional, on the topic of the benefits of digitization in academic practice, 59 respondents commented on the advantages of digitalization and 61 respondents on the disadvantages. At the FE TUL, 37 academic staff members commented on the advantages and disadvantages of digitalization.

Academics at both FE TUL and the selected German universities agree on the main benefits of digitalization in their work during the pandemic. However, according to the German respondents, the advantages of digitalization, such as flexibility in teaching and the choice of communication methods, ensuring the timeliness and availability of learning materials, improving the quality of teaching and independence from the location of students and teachers, are given twice as much importance. One in twelve respondents from the academic staff at FE TUL know the need to acquire new digital skills during the pandemic and perceive them as assets in their work. Also, one in twelve respondents working at FE TUL did not identify any benefits of digitalization in academic staff work, which was different for academic staff at the selected German university.

The main disadvantage that digitalization has brought to the work of academics at FE TUL and the selected German university is the lack of social contact with students and colleagues, which affects knowledge sharing in the workplace.

It should also be noted that one in six respondents from Czech and German academics experienced the need to spend more time preparing online teaching materials. In addition, one in ten respondents on both the Czech and German sides felt a need for more feedback and non-verbal communication with students during online teaching during the COVID-19 pandemic. Based on the results obtained, academics at the selected German university encountered declining student concentration and motivation in the online environment and technical problems in providing online teaching more often than at FE TUL.

In conclusion, the results of the international comparison confirm similar perceptions of the advantages and disadvantages associated with digitalization in academic work during the pandemic period.

References

1. Mazurchenko, A., Zelenka, M.: Employees' digital competency development in the construction and automotive industrial sectors. Cent. Eur. Bus. Rev. **11**, 41–63 (2022). https://doi.org/10.18267/j.cebr.284
2. Mazurchenko, A., Zelenka, M., Maršíková, K.: Demand for employees' digital skills in the context of Banking 4.0. E & M Ec. Man. 25, 41–58 (2022)

Chapter 7
Summary of Research Results, Discussion and Contributions

Abstract The book examined the impact of digitalization in the context of HRM at the theoretical and practical levels. From a theoretical perspective, Chaps. 2, 3, and 4 focused on defining basic concepts and comparing definitions in order to contribute to the definition of a theoretical framework of digitalization in HRM. The secondary data analysis presented in Chap. 5 mapped the main findings of recent global surveys on the current and desired level of employees' digital competencies. The empirical results are presented in Chap. 6 and include the results of a research investigation of digitalization in the context of HRM from the perspective of both employees and employers in Czech and foreign organizations operating in selected industries. Chapter 7 introduces the main research results and their discussion and presents the contributions of this book both in theoretical and practical terms. Chapter 7 also presents the limitations of the leading research in this book and suggestions for future research in the area of digitalization in HRM.

7.1 Summary of the Research Results

The theoretical part of the submitted book was based on a systematic literature search of scientific publications registered in the renowned WoS and Scopus databases. The literature search demonstrated the growing importance of digitalization for the academic community, as expressed by the increasing number of publications on this topic over the last 14 years.

It was also found that the current literature needs to take a uniform approach to defining terms related to digitalization, which is why the book's Author proposed her definition of digitalization to HRM.

The literature search verified the impact of industrial revolutions on HRM over the last century. It confirmed their influence on redefining the role of HR managers in the context of the ongoing socio-technical revolution.

In the theoretical part of the book, each author's perspective showed the main differences between traditional and digital HRM. It revealed the difference in global human capital trends before the pandemic and during COVID-19.

A. Mazurchenko, *The Impact of the Digitalization on the Human Resource Management*, Information Systems Engineering and Management 25,
https://doi.org/10.1007/978-3-031-76790-6_7

The results showed the impact of the COVID-19 pandemic on the role of HRM and the key benefits, risks, and challenges caused by digitalization in HRM. The literature search validated approaches to developing competency models and their potential use in HRM.

It also confirmed a greater emphasis on incorporating digital competencies into organizational competency models and the impacts of digitalization on the competence profiles of employees in different occupations.

Digitalization in HRM and the development of employees' digital competencies have been mapped in terms of theory in an international context. However, this topic is not addressed in Czech literature in connection with the ongoing COVID-19 pandemic, and the Author of this book aims to fill this gap in knowledge.

Based on the literature search, the Author of the book formulated five RQs:

- RQ_1 "What competencies are necessary for the effective use of DTs at work in the opinion of employers and employees?"
- RQ_2 "How do priority digital competencies differ across sectors?"
- RQ_3 "How often and how do employees acquire the digital competencies they need for their work?"
- RQ_4 "What are the main barriers to developing digital competencies for work from the perspective of employees and employers?"
- RQ_5 "How does digitalization affect organizational learning and development and competence models applied in organizations?"

A range of secondary data and research on the impact of digitalization on the labor market and employee competencies was also used to assess the topic. The Author of the book verified the skills requirements of employees for the future world of work and the competency requirements needed for HR professionals in the future and proposed a framework of digital competencies in the context of digitalization. It has also been shown that digitalization accelerated by the ongoing COVID-19 pandemic affects the industry differently.

Based on the results of the detailed secondary data analysis presented in Chap. 5, key sectors were selected for further exploration of the impact of digitalization on employee competencies in the empirical part of the book.

The actual research on digitalization took place between 2019 and 2022 and was based on this data collection:

- Pilot survey on "Digitalization and its impact on businesses and employee competencies."
- Research on "Digital competence in corporate practice."
- Research on "Digital readiness of current employees."
- Research on "Digital readiness of academics."

The pilot survey on "Digitalization and its impact on businesses and employee competencies" within the book's framework aimed at determining Czech employers' interest in digitalization and developing employees' digital competencies. The results of the conducted research confirmed the topicality of the researched topic, which is also emphasized by [1–3]. The conducted pilot research supported the views of

[1], who believe that successfully integrating DT into business processes requires systematic T&D of employees in digitalization.

Assumption 1 confirmed that businesses' use of DT for their day-to-day activities is independent of the size of these enterprises according to the number of employees. *Assumption 2, which is that* top management and employees' views on employees' readiness to use the new DT at work differ, was also confirmed. For *Assumption 3*, which sought to find out whether there is a statistically significant relationship between respondents' reactions to the selected statements, it can be concluded that employers who make more use of DT in production within their business are also more likely to rate employees' problem-solving skills as critical as a result of the progressive digitalization of the workplace (a relatively weak direct relationship was found here). Businesses that use DT most in production were not shown to be planning to invest in process and product innovation in the next five years.

On the contrary, it was confirmed that employers who consider employees' problem-solving skills necessary as a result of the progressive digitalization of the workplace are more likely to intend to invest in process and product innovation over the next five years (here, a relatively weak direct relationship was found) (see Sect. 6.1).

The survey on "Digital Competencies in Corporate Practice" aimed to map the current situation in the field of digitalization in HRM and identify Czech companies' needs in terms of the employees' digital competencies.

One of this book's main contributions is comparing the views of Czech enterprises in the construction, automotive, banking, and insurance sectors. This book provides the most up-to-date knowledge in the field of digitalization in HRM during the COVID-19 pandemic in the context of Czech enterprises in selected sectors. Based on the determined results, it can be argued that the digitalization process in the studied companies was greatly accelerated, which is also mentioned by [56].

It was not confirmed that the level of employees' digital competencies of Czech construction and automotive companies, banks, and insurance companies was statistically significant depending on the sector according to the CZ-NACE classification of economic activities (see H_1 in Sect. 6.1).

Similarly, it was not confirmed that the frequency of improving employees' digital competencies was statistically significant depending on the sector in which the employer operates (see H_2 in Sect. 6.1).

In contrast, a statistically significant dependency was determined on the level of readiness of Czech employers in the construction, automotive, banking, and insurance sectors to use DT in the context of the COVID-19 pandemic being dependent on the sector (see H_3 in Sect. 6.1). It is supported by the results of a study conducted by [40] in the CR, which highlights the different attitudes of employers towards digitalization in the context of the COVID-19 pandemic, depending on the sector.

The research on **"Digital Readiness of Current Employees"** was conducted in 2021 and focused on exploring employee attitudes toward digitalization and developing digital competencies during the COVID-19 pandemic.

The research on "Digital Readiness of Academic Staff" was conducted between 2021 and 2022 and aimed to assess the readiness of academic staff at FE TUL and a selected HEI in Germany during the coronavirus pandemic. This international comparison has not yet been carried out and is a unique and timely contribution to the present book.

Summary findings taking into account the results of all four surveys are presented in Table 7.1.

Based on the conducted analyses (i.e., RQ_1 and RQ_2), it can be concluded that, on the employers' side, representatives of the construction, automotive, banking, and insurance sectors in the CR agree on the top digital skills most needed by their employees. From the perspective of academics in the CR and Germany, the priority of digital skills is the same, given that in both countries, they were triggered by the need to move to online learning during the COVID-19 pandemic. Both interviewed employers and employees agree that basic data skills and the ability to cope with the

Table 7.1 Responses to the research questions with the opinion of employees and employers

	Employers			Staff		
	Automotive industry	Construction	Banking and insurance	Employees working in different sectors	Academic staff at FE TUL	Academics at a selected German university
RQ_1, RQ_2	Basic data entry and processing skills, data skills Problem-solving skills arising from working in a digital environment			Basic data skills and problem-solving skills in a digital environment	Digital skills for teaching, basic data entry and processing skills, complex data analysis skills, and mathematical skills	
RQ_3 Frequency of education	A third of Czech banks and insurance companies, a fifth of construction companies, and an eighth of automotive companies provide continuing education			X	Self-study, informal help from colleagues and friends, in-house training	Self-study, offers from the Centre for eLearning, offers from faculty, informal help from colleagues
RQ_3 Most common methods of education	In-house (internal) training is the preferred method in all three sectors					
RQ_4	Employee workload, resistance to accepting changes in the work environment, and learning new things in response to the COVID-19 pandemic			Lack of technical support at the workplace and the need to align the training course's content in digitalization with the employer's needs	Workload, lack of faculty/university support, lack of technical support in the workplace	Workload, lack of goals for further education, especially in digitalization, and other priorities

challenges of working in a digital environment are crucial in a pandemic, regardless of sector.

The above research results were supported by four in-depth interviews conducted in 2020 with representatives of three companies in the manufacturing industry and one large Czech bank. The results, elaborated in the form of four case studies, provide valuable insight into the impact of digitalization on HRM and employee education and development regarding the use of competency models in the COVID-19 pandemic.

The thesis's theoretical and empirical results, as well as the contributions in theoretical and practical terms, are described in more detail.

7.2 Theoretical Benefits of the Research

The book's main contribution in terms of theoretical knowledge is a comprehensive examination of selected aspects of the phenomenon of digitalization in the field of HRM and the development of employees' digital competencies during the COVID-19 pandemic. This showed the main trends in this area and, in the context of the turbulent developments in recent years, provided a deeper insight into identifying the benefits and risks associated with digitalization, both from the perspective of employees and employers. Given the unpredictable development of the coronavirus pandemic, Czech professional literature sources need to sufficiently cover this topical issue. The findings presented in the systematic literature review in Chap. 3, 4, and 5 significantly expand the field of science.

Considering that the current Czech and foreign scientific literature does not uniformly define digitalization and that several definitions develop the concept and complement each other, the Author of this book proposed her comprehensive definition of digitalization in relation to HRM (see Sect. 2.2).

The presentation of a historical perspective on the development of HRM under the influence of all five industrial revolutions is also a contribution on the theoretical level. The Fifth Industrial Revolution began in 2021 during the coronavirus pandemic, and its impact on HRM has yet to receive much attention in the Czech scientific and professional literature (see Sect. 3.1).

This book identified the main components of competence based on definitions of this concept proposed by Czech and foreign authors over the last 40 years (see Sect. 4.1).

The theoretical contribution also includes a detailed mapping of competency models emphasizing verifying the need and position of digital competencies in the context of different organizations in the CR and abroad (see Sect. 5.2). It can also contribute to theoretically defining the process of developing a competence model, which can inspire organizations to consider introducing a competence model. Trexima last surveyed the use of competency models in Czech companies as part of the fourth wave of the HR Monitor in 2014. This points to a gap in theoretical knowledge concerning the practical use of competence models in the context of the

Fourth Industrial Revolution during the COVID-19 pandemic (see Sect. 5.2). The literature review also confirmed the relevance of the issue and the need to include digital competencies in the competence models of organizations.

7.3 Benefits of the Research in Practical Terms

The practical contribution of the present book is the acquisition of current primary data that enabled the assessment of the state of digitalization in managing leadership resources and developing employee competence profiles. Thanks to these findings, an analysis of the current state-of-the-art and a comprehensive assessment of the perspectives of employees and employers was made on the issue of digitalization and T&D in employees' digital competencies during the pandemic period.

Another contribution is that the empirical part of the book provides an international comparison of the development of digitalization, the level of digital competencies, and the use of tools in the education and development of academic staff using primary data from Czech and German HEIs. In the case of the investigation of the digital readiness of employees, this is the first such research at both a Czech and German university; moreover, in the case of the German university, processing in both English and German was ensured.

The book's practical contribution can also be used to prepare the final report and present the results to the university management. The Author formulated recommendations for managing a German HEI, emphasizing the need for further T&D of academic staff in digital competencies.

The practical benefits of the four surveys presented in Chap. 7 can be summarized in the following recommendations for employers:

- Integrate digital competencies into the competence models applied in organizations and intensify the use of this approach;
- Promote a culture of continuous and systematic learning and development in the workplace by having employers set aside time during the work week for employees to participate in digital learning activities and actively supporting those who are willing to learn and develop their digital competencies;
- Encourage informal knowledge and experience sharing among staff on the use of digital tools and DT in the workplace;
- Inclusion of data skills courses in the training programme (possibly expanding the offer to include advanced courses and training), given that both employers and employees consider this area of T&D to be a priority;
- Expanding the digital infrastructure of businesses and improving teleworking opportunities to promote hybrid forms of work where the specificities of the sector and the employer's operations allow it;
- Increase HRs in IT departments of companies, as their workload increased significantly during the COVID-19 pandemic, and employees began to feel a lack of

individual support in dealing with the challenges of using the new DT in their work.

The book also contributes to four case studies that reflect the impact of digitalization on the training and competence profiles of employees in three Czech industrial companies and one sizeable Czech bank during the COVID-19 pandemic.

7.4 Limits of Empirical Research

Even though the empirical research presented in this book provides valuable insight into the topic under investigation during the COVID-19 pandemic, it has several significant limitations:

- The pilot research was based on the results obtained from a limited number of respondents who were also interested in digitalization and its impact on business practice. Thus, their perception and knowledge of the topic may be greater than the average respondent's.
- A limitation is the low return rate of questionnaires (i.e., 2.0%) when investigating the phenomenon of digitalization and the development of digital competencies from employers' perspectives, which fails to ensure the representativeness of the results obtained.
- The extension of the empirical research was carried out on selected sectors of the economy (i.e., the manufacturing industry with emphasis on the automotive industry, construction, banking, and insurance, and the education sector with an emphasis on HEIs). Thus, the results obtained from the primary data cannot be generalized to the entire population; however, within HEIs, for example, these institutions can interpret the results.
- The book's Author developed only a limited number of case studies reflecting the actual situation in the field of digitalization in HRM in the CR. These were three case studies of Czech automotive companies and one sizeable Czech bank on the impact of digitalization during the COVID-19 pandemic.
- Exploring the digital readiness of employees is limited by the small research sample (i.e., 31 responses from respondents working in different industries).
- The research on digitalization and development of digital competencies from employees' perspective only covers a limited geographical area (i.e., only five regions of the CR).
- The survey does not consider the views of employees under 20 and over 50. The respondents were only people aged 20–50 years because they were students of the CS at FE TUL.
- The primary data collection from employees did not consider their job position, limiting a more detailed assessment.
- The survey of academic staff was only carried out in two selected universities (one Czech and one German university, whose management was interested in the uniqueness of the research results) , and the results obtained cannot be generalized

beyond the institutions in question. Nevertheless, the possibility of comparison in an international context in this setting is a significant contribution of the book.

Despite the limitations mentioned above, the presented book significantly complements and extends the existing knowledge on the issue of digitalization in HRM that has been investigated.

7.5 Suggestions for Future Research on Digitalization in Human Resource Management

Based on the literature search and primary research conducted in this book, several directions for future in-depth research can be suggested:

- Future work could focus on exploring the impact of digitalization in sectors of the economy other than automotive, construction, banking and insurance.
- Future research could be extended to other countries outside the CR.
- Further research is desirable to explore employees' digital readiness concerning their specific job roles or the department in which they work within companies.
- A continuation of the research on employees' digital readiness could focus on the age group below 20 years and above 50 years, thus complementing and extending the results of the empirical research already conducted.
- In the future, it would be possible to focus on examining the digital readiness of employees across TUL and make cross-faculty comparisons.
- Further research is also needed to explore the digitalization's impact on HR professionals' competence profiles.
- It would be helpful to explore ways to reduce digitalization's impact on business practice and increase employees' digital readiness in response.
- Continuing research on employee digital readiness can explore how to most effectively train and develop employees in digital competencies and what methods to use.

The above themes for future research show that digitalization in HRM and the development of employees' digital competencies are pervasive and topical, giving the Author of the book room for further research.

References

1. Balsmeier, B., Woerter, M.: Is this time different? How digitalization influences job creation and destruction. Res. Pol. **48**, 1–10 (2019). https://doi.org/10.1016/j.respol.2019.03.010
2. Janssen, J., Stoyanov, S., Ferrari, A., Punie, Y., Pannekeet, K., Spoel, P.: Experts' views on digital competence: Commonalities and differences. Comp. Ed. **68**, 473–481 (2013). https://doi.org/10.1016/j.compedu.2013.06.008

3. Oberländer, M., Beinicke, A., Bipp, T.: Digital competencies: A review of the literature and applications in the workplace. Comp. Ed. **146**, 103752 (2020). https://doi.org/10.1016/j.compedu.2019.103752

The manufacturer's authorised representative in the EU is Springer Nature Customer Service Centre GmbH, Europaplatz 3, 69115 Heidelberg, Germany. If you have any concerns regarding our products, please contact ProductSafety@springernature.com

Printed and bound by CPI Group (UK) Ltd, Croydon, CR0 4YY
01/10/2026
02225839-0008